Low-Switching Frequency Modulation Schemes for Multi-level Inverters

Low-Switching Frequency Modulation Schemes for Multi-level Inverters

A. Rakesh Kumar
T. Deepa
Sanjeevikumar Padmanaban
Jens Bo Holm-Nielsen

CRC Press
Taylor & Francis Group
Boca Raton London New York

CRC Press is an imprint of the
Taylor & Francis Group, an **informa** business

First edition published 2021
by CRC Press
6000 Broken Sound Parkway NW, Suite 300, Boca Raton, FL 33487-2742

and by CRC Press
4 Park Square, Milton Park, Abingdon, Oxon, OX14 4RN

First edition published by CRC Press 2021

CRC Press is an imprint of Taylor & Francis Group, an Informa business

ISBN 13: 978-0-367-55260-2 (pbk)
ISBN 13: 978-0-367-51290-3 (hbk)
ISBN 13: 978-1-00-305319-4 (ebk)

Typeset in Computer Modern font
by KnowledgeWorks Global Ltd.

To the Masters of Sahaj Marg
(The Heartfulness Movement)

Contents

Contents

Foreword

It is my pleasure to write this foreword for the book entitled, "Low-Switching Frequency Modulation Schemes for Multilevel Inverters". A quick look at the book was sufficient enough for me to understand that it addresses the academicians and industrialists to be an attractive book. It has a simple approach to delivering the concept of new Cross-Connected MLI, symmetrical/asymmetrical configurations of H-bridge, and Multilevel DC link inverters with different novel low switching-frequency modulation methods.

Power electronics play an essential role in power conversion; multilevel inverters are matured field and have found numerous applications in industrial power sectors. The book has found a way to present cutting edge research finding for new multilevel converters, which generates high-quality output power with limited noise (Total Harmonic Distortion). Further, the authors have shown clearly dedication with the numerical development in Matlab/Simulink software with theoretical predictions, and laboratory-scale experimental prototype verification. In fact, the performance of low switching frequency schemes is still to be exploited for its full potential has thrown a research area wide open to all! It is especially true with the renewable energy systems (PV, Wind, etc.), these investigation results provide comprehensive solutions and for whole power energy society.

This book is a collaborative effort between Vellore Institute of Technology, Chennai, India, and Aalborg University, Esbjerg, Denmark, on the joint venture of close operation and technological sharing between researchers. I congratulate the authors in providing a new dimension to this rapidly progressing research area of the power converter.

I am sure that all readers will enjoy this book for its technical contribution and sufficient to gain knowledge to take up research in this exciting field of Power Electronics—Multilevel Converters and Low-Frequency modulation methods.

Finally, I congratulate the Authors and the CRC publication for the effort to make this book outcomes as technical exploration for the scientific society.

Good Luck!

Prof. Frede Blaabjerg, Fellow IEEE
Villum Investigator Full Professor
Center of Reliable Power Electronics (CORPE)
Department of Energy Technology
Aalborg University, Denmark

Preface

Multilevel Inverters (MLIs) are power electronics converters in the field of electrical engineering used for the conversion of DC power to ac power. The MLIs are a revolution in the DC to ac power conversion replacing the traditional two and three level inverters. They are one of the fast evolving research topics in power electronics. Hence, this book addresses the academicians as well as researchers alike.

This book is divided into six chapters. The first chapter deals with the classical MLI topologies and Modular Multilevel Converters. Modular Multilevel Converters are a new breed of MLIs which have gained sufficient attention in the present. The second chapter is dedicated to different High Switching Frequency (HSF) and Low Switching Frequency (LSF) modulation schemes. The various advantages of LSF over HSF modulation schemes are also discussed. A brief comparison of the Nearest Level Modulation (NLM) scheme with other LSF schemes presenting the benefits of NLM is also shown. Towards the end, a recently introduced modified version of NLM scheme is presented which adds to the family of LSF modulation schemes. The third chapter presents the implementation of various LSF modulation schemes on Cross-Connected Sources based MLI. The simulation and experimental results are also presented. Similarly, the fourth ad fifth chapter deals with the implementation of LSF modulation schemes on Cascaded H-Bridge MLI and Multilevel DC-Link Inverter respectively. The last and sixth chapter provides an insight into some of the practical application of MLI in power conversion of Renewable Energy Sources (RES). The future scope of LSF modulation schemes for RES application are discussed towards the end.

The book is of significant value to researcher as the algorithm for modified Nearest Level Modulation Scheme will provide the necessary input to carry out further research in improving the performance of the low switching frequency modulation schemes.

This book will serve the academic and research community alike. MLI is an emerging research area and hence, proper knowledge needs to be provided to equip the academic community well. The book focuses especially on low-switching frequency modulation schemes which over numerous advantages over the high-switching frequency modulation schemes.

Author Biography

A. Rakesh Kumar completed his B.E (Honors) in Electrical and Electronics Engineering from DMI College of Engineering, Anna University, Chennai, India in 2011 and M.Tech in Power Electronics and Drives from Jerusalem College of Engineering, Anna University, Chennai, India in 2013. He worked as Assistant Professor with the Department of EEE, Rajalakshmi Engineering College, Chennai, India from 2013 to 2015. He received his PhD degree from Vellore Institute of Technology (VIT), Chennai Campus in 2020. He was working as Teaching cum Research Assistant from 2015 to 2019 at VIT, Chennai. Currently, he is a Post Doctoral Fellow with the Department of EEE, National Institute of Technology, Tiruchirappalli, India. He is a member of IEEE. His field of interest includes multilevel inverters, inverter modulation techniques, smart grid and its applications. (E-mail: rakesh.a@ieee.org)

T. Deepa is currently working as Associate Professor in School of Electrical Engineering from Vellore Institute of Technology, Chennai Campus. She completed her PhD from College of Engineering, Guindy, Anna University in Process Control. She completed her B.Tech in Electrical and Electronics Engineering from Manonmaniam Sundaranar University, Tirunelveli and M.Tech from College of Engineering, Guindy, Anna University in Power System. Her research interest includes process control, control systems and intelligent controllers. (E-mail: deepa.t@vit.ac.in)

Sanjeevikumar Padmanaban (M'12, SM'15) received the bachelor's degree in electrical engineering from the University of Madras, India, in 2002, the master's degree (Hons.) in electrical engineering from Pondicherry University, India, in 2006, and the PhD degree in electrical engineering from the University of Bologna, Italy, in 2012. He was an Associate Professor with VIT University from 2012 to 2013. In 2013, he joined the National Institute of Technology, India, as a Faculty Member. In 2014, he was invited as a visiting researcher with the Department of Electrical Engineering, Qatar University, Qatar, funded by the Qatar National Research Foundation (Government of Qatar). He continued his research activities with the Dublin Institute of Technology, Ireland, in 2014. He was an Associate Professor with the Department of Electrical and Electronics Engineering, University of Johannesburg, South Africa, from 2016 to 2018. Since 2018, he has been a Faculty Member with the Department of Energy Technology, Aalborg University, Esbjerg, Denmark. He has authored 300 plus scientific papers and has received the Best Paper cum Most Excellence Research Paper Award from IET-SEISCON'13, IET-CEAT'16 and five best paper award from ETAEERE'16 sponsored Lecture note in Electrical Engineering, Springer book series. He is a fellow Institution of Engineers (FIE'18, India), fellow Institution of Telecommunication and Electronics Engineers (FIETE'18, India) and Fellow the Institution of Engineering and Technology (FIET'19, UK). He serves as an Editor/Associate Editor/Editorial Board of refereed journal, in particular, the IEEE Systems Journal, the IEEE Transactions on Industry Applications, the IEEE Access Journal, the IET Power Electronics, Wiley-International Transactions on Electrical Energy Systems, and the subject editor of the subject Editor of IET Renewable Power Generation, the subject Editor of IET Generation, Transmission and Distribution, and the subject editor of FACTS journal, Canada. (Email: san@et.aau.dk)

Jens Bo Holm-Nielsen currently works at the Department of Energy Technology, Aalborg University and Head of the Esbjerg Energy Section. On this research, activities established the Center for Bio-energy and Green Engineering in 2009 and serve as the Head of the research group. He has vast experience in the field of Bio-refinery concepts and Bio-gas production–Anaerobic Digestion. Implementation projects of Bio-energy systems in Denmark with provinces and European states. He served as the technical advisory for many industries in this field. He has executed many large scale European Union and United Nation projects in research aspects of Bio-energy, bio refinery processes, the full chain of bio-gas and Green Engineering. He has authored more than 100 scientific papers. He was a member on invitation with various capacities in the committee for over 250 various international conferences and Organizer of international conferences, workshops and training programmes in Europe, Central Asia and China. Focus areas Renewable Energy—Sustainability—Green jobs for all. (Email: jhn@et.aau.dk)

Symbols

Symbol Description

m	Number of MLI levels	N_{source}	Number of DC sources
α	Switching angle	V	Voltage
V_{DC}	DC Voltage Source	A	Ampere
j	Number of input DC voltage source	W	Watt
		Hz	Hertz
N_{level}	Number of MLI levels	V_{rms}	rms voltage
N_{DC}	Total number of DC sources		
N_{switch}	Total number of switches	x	Number of full bridge inverter
V_{max}	Magnitude of maximum voltage	ϕ	Phase

1

Introduction

A Multilevel Inverter (MLI) is a specialized field of power electronics field concerned with the conversion of multiple DC sources into a staircase-type AC waveform. There are several MLI configurations available, each employing a different number of semiconductor devices for achieving various MLI levels. Even though there are conventional two-level inverters to convert DC to AC power, MLIs are employed for their ability to generate AC output with less harmonic distortion. The MLIs are also vested with the ability to draw low distortion input current from the input DC sources [1].

An MLI modulation scheme refers to the change in the state of the power semiconductor switches from one operating mode to another operating mode. The modulation scheme can be a High-Switching Frequency (HSF) or a Low-Switching Frequency (LSF) modulation scheme [2]. An HSF modulation scheme refers to the operation of semiconductor switches above 1 kHz while an LSF modulation scheme refers to an operation of Insulated-Gate Bipolar Transistor (IGBT) switches below 1 kHz. The duty cycle of switching pulses to semiconductor switches play a crucial role. There are variations in the performance of the modulation schemes depending on it.

There are various kinds of modulation schemes with their employment depending upon the applications. Enormous research work has been carried out on HSF modulation schemes. Some of the HSF modulation schemes are Sine Pulse Width Modulation (SPWM), Selective Harmonics Elimination (SHE), Space Vector Control (SVC) etc. There are also LSF counterparts for the same modulation schemes [3].

In an HSF modulation scheme, a reference signal and a high-frequency carrier signal are compared. This leads to the generation of switching pulses. The same applies to an LSF modulation scheme with variation in the frequency of the carrier signal.

The rise of multilevel conversion is seen as a welcome revolution in the era of industrialization. The industrial era runs on high and medium voltage but high power applications. Hence, it is necessary to employ multiple semiconductor switches rather than a single semiconductor switch. The drawbacks of conventional two-level inverters are HSF modulation scheme, poor output voltage waveform, higher dv/dt stress, higher electromagnetic interference, need of LC filter etc., [4].

Employing an increased number of semiconductor switches is well suitable for high power conversion. The semiconductor switches are thus allowed to share the voltage and current through the converter. This helps in increasing the life of the converter. The multilevel power conversion has found itself demanding with the advent of renewable energy sources such as solar energy, wind energy [5–7]. MLIs with the convenience of adding isolated and individual sources is well suitable for integrating different types of renewable energy sources.

The MLIs have found widespread applications. Some of the important applications are:

1. Recently MLIs have found use in the control of variable speed drives and railway transportation electrification. The performance of induction motor drive is enhanced using intelligent rotor resistance estimator [8–10].

2. The research carried out in [11,12] have demonstrated that MLIs are employed in renewable energy systems, especially with PV systems for the conversion of the DC input from solar panel to AC output.

3. In [13, 14], MLIs were applied in the field of power systems and power electronics in STATCOM and induction heater system.

1.1   Classical Multilevel Inverter (C-MLI) Topologies

This section presents the three conventional MLI topologies along with a brief explanation of each topology.

1.1.1   Diode-Clamped MLI

The first MLI topology was invented by Nabae et al. in the year 1981, as shown in Figure 1.1. It was named as a neutral-point-clamped inverter and later also familiarized itself as Diode-Clamped MLI (DC-MLI). The DC-MLI is a combination of two DC sources, two diodes and four switches for a three-level AC output. The modes of operation are shown in Table 1.1. The levels can be extended further with an increase in the number of sources, diodes and switches. The diodes handle the circulating current in DC-MLI at the instant of zero-level.

The advantages associated with a DC-MLI is that since all the phases carry share a common DC bus, the need for the number of capacitors is also reduced. This also enables the back-to-back topology connection for high-voltage and drives applications.

The DC-MLI also operated efficiently with an LSF modulation scheme when the number of levels is increased. But as the number of levels increases, the diode requirement also increased, and this makes the circuit heavier [15].

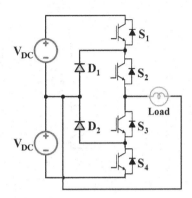

FIGURE 1.1: Diode Clamped MLI Topology

TABLE 1.1: Modes of Operation of Diode Clamped MLI Topology [4]

Level	Status of Switch S_1	Status of Switch S_2	Status of Switch S_3	Status of Switch S_4
0	OFF	ON	ON	OFF
V_{DC}	ON	ON	OFF	OFF
0	OFF	ON	ON	OFF
0	OFF	ON	ON	OFF
$-V_{DC}$	OFF	OFF	ON	ON
0	OFF	ON	ON	OFF

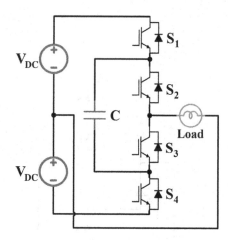

FIGURE 1.2: Flying Capacitor-MLI Topology

1.1.2 Flying Capacitor-MLI

The Flying Capacitor-MLI (FC-MLI) emerged from the DC-MLI by replacing the diodes with capacitors [16] to overcome the disadvantages of DC-MLI. Figure 1.2 shows the three-level FC-MLI where two DC sources, four switches and one capacitor is used. The modes of operation of the same are shown in Table 1.2. At the instant of zero-level, the capacitor is used to stabilize the MLI.

TABLE 1.2: Modes of Operation of Flying Capacitor-MLI Topology [4]

Level	Status of Switch S_1	Status of Switch S_2	Status of Switch S_3	Status of Switch S_4
0	OFF	ON	ON	OFF
V_{DC}	ON	ON	OFF	OFF
0	OFF	ON	ON	OFF
0	OFF	ON	ON	OFF
$-V_{DC}$	OFF	OFF	ON	ON
0	OFF	ON	ON	OFF

The capacitors used in the FC-MLI allows for the controlled flow of real and reactive power. Phase redundancy is another feature of the FC-MLI to enable capacitor voltage balancing. But like the DC-MLI, as the number of levels increases, the need for capacitors also increase contributions to a higher cost and heavier circuit [15].

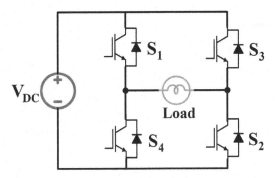

FIGURE 1.3: Three-Level H-bridge Topology

1.1.3   H-Bridge MLI

The H-Bridge is third in the list of conventional MLI topologies shown in Figure 1.3, which can generate a three-level output [17]. Table 1.3 shows the status of the different switches to generate a three-level output. A Cascaded H-Bridge (CHB) inverter is a series combination of several individual H-bridge Inverter to achieve a higher number of levels. The H-bridge inverter does not consist of any diode or capacitor, which makes it superior to the other two conventional MLI topologies.

TABLE 1.3: Modes of Operation of Single H-Bridge [4]

Level	Status of Switch S_1	Status of Switch S_2	Status of Switch S_3	Status of Switch S_4
0	ON	OFF	ON	OFF
V_{DC}	ON	ON	OFF	OFF
0	ON	OFF	ON	OFF
0	OFF	ON	OFF	ON
$-V_{DC}$	OFF	OFF	ON	ON
0	OFF	ON	OFF	ON

The CHB inverter is highly modular in structure. This means that it is easy to add or deduct a small subset of the MLI structure as per the requirement of the topology. The CHB inverter also paves the way for being integrated with other MLI topologies to form a new class of MLI topologies. There have been seemingly a more significant number of applications in industries than DC-MLI and FC-MLI because of its advantages.

1.2 Modular Multilevel Converters

1.2.1 Introduction to Topology

The Modular Multilevel Converters (MMC) are the most recent advancements in the field of MLI topologies. The features of MMCs are transformer-less operation, redundant topology, modular and extendible structure.

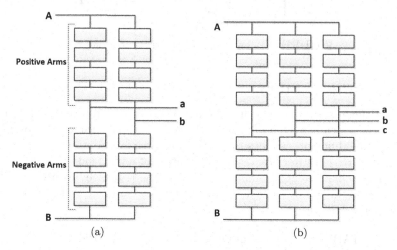

FIGURE 1.4: Modular Multilevel Converter Topologies. (a) Single Phase Topologies; (b) Three Phase Topologies (each box represents a single power cell) [18]

The general structure of an MMC is shown in Figure 1.4. There are many combinations in which the power cells can be combined with each other. The combination of cells is categorized into positive and negative arms, as shown in Figure 1.4. An MMC is a combination of several power cells which perform a cascaded operation. These power cells are realized in the form of smaller power electronics converter, as shown in Figure 1.5. The power cells can be in the form of a single H-bridge, half-bridge, unidirectional half-bridge, DC-MLI or FC-MLI. A single H-bridge or a half-bridge is most often used in an MMC because of fewer components and improved control circuitry.

The operation of an MMC requires several means of controlling the voltage or current. One way of controlling the output current of MMC is to control the individual arm currents. A reference current is to be taken while controlling the current. For this purpose, a closed-loop control strategy can be implemented. Due to the presence of several power cells, a circulating current is always flowing through the power cells. The power cells form a loop within itself and giving rise to circulating currents. By choosing a proper modulation index and resonant converters, the circulating currents can be reduced.

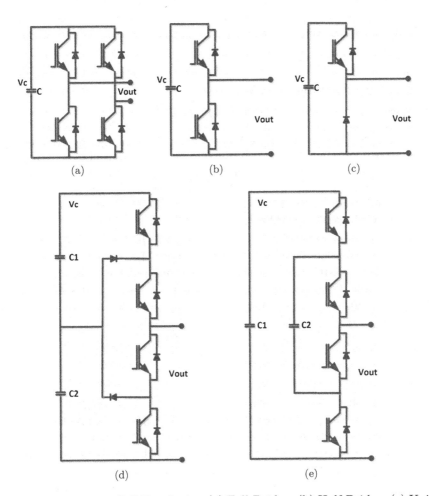

FIGURE 1.5: Power Cell Topologies. (a) Full-Bridge; (b) Half-Bridge; (c) Unidirectional Half-Bridge; (d) DC-MLI and (e) FC-MLI [18]

Capacitor voltage control is an essential function in the operation of MMC. Each power cell in an MMC is considered to be equivalent to a voltage source with the use of a capacitor. A reference voltage can be taken to balance the capacitor voltage by varying the modulation index of the PWM signals. Apart from these controls, the MMC as a whole can be operated with HSF or LSF modulation schemes. Triangular waveform based carrier signals can be used with techniques such as phase-shifting and level shifting. Space vector modulation schemes are another way of operating the MMC under high switching frequency. Under the LSF modulation schemes, Selective Harmonics Elimination and Nearest Level Modulation schemes are well-known techniques. The NLM scheme, in particular, is a straightforward method and highly suitable for an increased number of MLI levels. SHE is also a suitable scheme, but with an increasing number of levels, the calculation of switching angles is complicated.

1.2.2 Applications of MMC

The MMCs have been used in the following applications [18]:

1. *HVDC Applications*: High Voltage DC transmission is given importance for long-distance transmission than AC transmission system because of the higher efficiency. The HVDC system is become more and more possible because of the advancements in the power electronics industry. For long-distance transmission, these systems offer excellent control and lesser EMI.

2. *Integration of Renewable Energy Sources (RES)*: Solar and wind energy can be seamlessly integrated with MLI topologies. This is because MMCs offer both isolated and non-isolated configurations. For solar energy, PV panels can be integrated with MMCs with isolated sources. PV modules can be added in series or parallel, and therefore they provide excellent flexibility in choosing the power rating and voltage rating of MMCs. For wind energy, DC-MLI with symmetric sources is currently investigated with various control schemes.

3. *Battery Energy Storage System*: Several battery banks can be added in parallel to form a Battery Energy Storage System (BESS). BESS forms as an input to the MLI topology and offers an excellent way of storing the excess of power when supply exceeds demand. The same BESS can also supply to the load with a bidirectional converter and meeting the load demand. For this purpose, MMC has a highly modular structure and much suitable for storage applications.

4. *Drives applications*: MMCs have found suitable applications for medium voltage drives. As the MMCs are modular in structure, the cells can be easily attached or removed to existing converters. This leads to a more straightforward construction of power electronics converters and scalable voltage ratings. The need for any output filter for standard line motors is also removed with the use of MMCs. The dynamic characteristics of MMCs can also be improved with drives application with the use of HSF modulation schemes.

2

Low Switching Frequency Modulation Schemes

The operation of MLIs requires giving proper switching pulses to ensure efficient operation of MLI. The modulation scheme essentially defines, how the transition in MLI takes place from one mode to another mode of operation. It is also essential to have dead time between each transition from one mode to another mode. This is due to the high possibility of the semiconductor switches getting short-circuited, in the absence of dead time.

A reference signal compared with a carrier signal leads to the generation of switching pulses. The reference signal is most often a sine waveform and compared with a carrier signal which, in most cases, is a triangular waveform. The output of MLI is desired to be in sinusoidal. Hence, a sine wave is the choice of the reference signal. However, the frequency of the AC output of MLI is maintained at line frequency unless there is any need for high-frequency AC output.

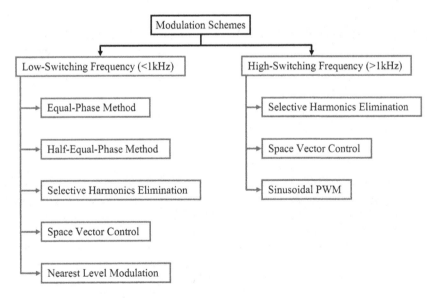

FIGURE 2.1: Classification of Modulation Schemes

There are many conventional modulation schemes implemented on MLIs. For the ease of simplicity and clear understanding, they can be classified into LSF and HSF modulation schemes. The LSF modulation schemes normally operate with a switching frequency of less than 1 kHz while the HSF modulation schemes operate with a switching frequency of more than 1 kHz. There are many modulation schemes available within LSF and HSF modulation schemes. Some of the modulation schemes can be operated in both LSF and HSF schemes. The classification of conventional modulation schemes is shown in Fig. 2.1. A brief discussion on each of this modulation scheme is discussed as follows.

2.1   High Switching Frequency (HSF) Modulation Schemes

1. *Selective Harmonics Elimination (SHE)*: The SHE scheme primarily takes care of reducing the lower order harmonics in the MLI output and thereby, offering a better AC waveform output quality. SHE can be implemented with high switching frequency where a reference signal is compared with high-frequency carrier signal to generate gate pulses. Recent research carried out have used different ways of implementing SHE with the use of innovative carrier and reference signals [19–21].

2. *Space Vector PWM (SVPWM)*: The gate pulses under SVPWM are generated based on the space vector coordinates. The efficiency of this scheme is based on the minimization of space errors between the vector coordinated and attaining the better quality AC output waveform. One significant drawback of this scheme is that for a higher number of levels, this scheme is complicated and results in too much of computation time [22–24].

3. *Sinusoidal PWM (SPWM)*: One of the most widely used HSF modulation schemes is the SPWM. Because an MLI output is ideally a sinusoidal output, therefore, it is preferred to generate the switching pulses which are close to the sinusoidal nature of AC output. A sine wave acting as a reference signal is compared with triangular carrier wave signal for the generation of gate pulses. This scheme offers ample innovation, and there has been a good amount of research carried out on semi-sine and quarter-sine waveforms. There are different variants of the SPWM known as Phase Dissipation (PD), Phase Opposition Dissipation (POD) and Alternate Phase Opposition Dissipation (APOD) [25–27].

2.2 Low Switching Frequency (LSF) Modulation Schemes

1. *Equal-Phase Method (EPM)*: EPM is a primitive modulation scheme where the pulse width of each level in the MLI levels are equal. This leads to a triangular waveform and carries a high amount of THD. The equation to arrive at the firing angles can be calculated using Eqn. (2.1).

$$\alpha_i = i\left(\frac{180°}{m}\right); \quad for \ \ i = 0, 1, 2, 3, ..., \frac{m-1}{2} \qquad (2.1)$$

2. *Half-Equal-Phase Method (HEPM)*: HEPM is another primitive modulation scheme where all the levels of the MLI carry equal pulse width but, for the pulse width of the highest level, it is double the pulse width of other levels. This also leads to a triangular waveform and somewhat better than EPM but carries a high amount of THD. The equation to arrive at the firing angles can be calculated using Eqn. (2.2).

$$\alpha_i = i\left(\frac{180°}{m+1}\right); \quad for \ \ i = 0, 1, 2, 3, ..., \frac{m-1}{2} \qquad (2.2)$$

3. *Selective Harmonics Elimination (SHE)*: SHE is found to be highly effective under the LSF modulation scheme. With a specific set of switching angles, it is often more natural and effective to eliminate the lower order harmonics. By using the standard Fourier equations set, it is possible to remove the required odd harmonics and improve the power quality of MLI output. SHE can also be implemented without the need for a closed-loop controller and hence suitable for various open-loop applications [28–30].

4. *Space Vector Control (SVC)*: Space vector theory is implemented in the SVC on the MLI applications. The space vector coordinates are used to determine the status of MLI switches. The space vector coordinates increase in number as the MLI level increases. This gives rise to complex operations. However, for MLI levels that are more than 13-level, SVC is still considered to be better as it offers a smooth transition from one state to the other [17, 31].

5. *Nearest Level Modulation (NLM)*: NLM, also referred to as Nearest Level Control (NLC) or Sine Property (SP), is a straightforward approach in determining the gate pulses to MLI. A comparison of the sine wave with a constant to generate gate pulses at exactly half of the magnitude of the rising edge of an MLI level will generate

an ideal MLI output. The equation 2.3 can be used to generate switching pulses to be fed to MLI switches [13, 32–34].

$$\alpha_i = sin^{-1}\left(\frac{i - 0.5}{n}\right); \quad for \quad i = 1, 2, 3, ..., n \qquad (2.3)$$

where, $n = \frac{m-1}{2}$

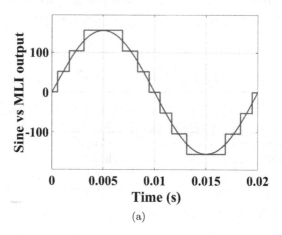

(a)

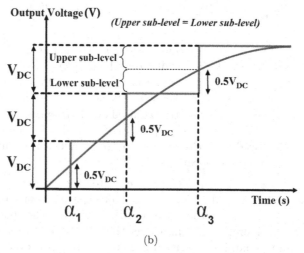

(b)

FIGURE 2.2: (a) A Typical NLM based Output Waveform for 7-Level MLI Output and (b) Enlarged View of NLM (A Quarter of the Full Cycle) [35]

2.3 Advantages of LSF over HSF Modulation Scheme

The LSF modulation schemes have many advantages over the HSF modulation schemes. They are listed as follows:

1. The LSF modulation scheme offers a reduced device switching stress which is a striking feature. Since the semiconductor switch operates with reduced frequency, it provides enough time to recover from the on and off switching states.

2. The device switching losses under LSF modulation scheme are lower due to the reduced stress. The voltage and current waveforms during on and off switching periods are improved under low frequency, and hence the power loss is also less.

3. For an HSF modulation scheme, the semiconductor switch generates enough heat as the frequency of operation is high, and hence, it becomes necessary to provide proper cooling. This is avoided when the semiconductor switch is operated under a lower frequency.

4. When operating under LSF modulation scheme, there is no special attention required for manufacturing the semiconductor and some relaxation can be given in the material and design used. This is due to reduced stress and reduced losses.

5. Since the losses are less for a semiconductor device under LSF modulation scheme, the input and output power from the device is more or less the same. Hence, there is a high device utilization.

6. In a power electronics converter, there are many devices, especially MLIs, where the device count goes to as high as 20. Hence, when the loss of each device is less, this leads to the overall improvement in the converter efficiency.

TABLE 2.1: Advantages of LSF over HSF Modulation Schemes

Sl.	Performance Parameters	HSF Schemes	LSF Schemes
1	Device switching stress	high	low
2	Device switching losses	high	low
3	Device cooling requirements	high	low
4	Device manufacturing cost	high	low
5	Device utilization	low	high
6	Converter efficiency	low	high

2.4   Advantages of Nearest Level Modulation Scheme over other LSF Modulation Scheme

The NLM scheme has excellent advantages over the other LSF modulation schemes. The drawbacks of all the LSF modulation schemes and the advantages of NLM scheme are discussed as follows [35]:

1. Equal-Phase Method (EPM) and Half-Equal-Phase Method (HEPM):

   (a) The EPM and HEPM schemes carry equal duty cycles and, hence, they yield a very poor voltage and current THD.

   (b) Since the THD is very high, they offer poor power conversion efficiency, and hence, they are not a recommended PWM scheme.

2. Selective Harmonics Elimination (SHE):

   (a) SHE is a useful scheme for the elimination of the lower order harmonics, but it comes at the cost of increased higher-order harmonics.

   (b) Any improvement in the SHE scheme requires an increase in the switching frequency, which is not desirable.

   (c) The SHE scheme is also highly dependent on the modulation index, and hence, for getting the maximum efficiency of SHE scheme, it is crucial to arrive at the exact modulation index.

3. Space Vector Control (SVC):

   (a) For a lower number of MLI levels, the working principles of SVC and NLM are almost the same. However, when the MLI levels increases, the SVC becomes increasingly complex.

   (b) The SVC scheme is ideal for use when a particular switching pattern is necessary concerning varying input voltages.

4. Nearest Level Modulation (NLM):

   (a) The striking feature of NLM is the simplicity in its execution. A simple equation is enough to find out the switching angles and thereby reducing the complexity of calculation.

   (b) With the NLM equation, it is straightforward to extend it to any number of MLI levels.

   (c) There is enough scope for improving NLM without increasing the switching frequency, which is the objective of the thesis.

The Table 2.2 shows the recent works carried out in the field of NLM scheme. A majority of the research work is on the use of conventional NLM scheme on conventional and new MLI topologies. There have been few advancements carried out to improve the NLM scheme. But all the advanced

NLM scheme offer a specific feature and not applicable to generalized MLI topologies. In the field of renewable energy applications, only the conventional NLM scheme has been implemented. Hence, there is high scope of the use of any modified NLM scheme with MLIs for renewable energy applications.

TABLE 2.2: Highlights of Recent Trends in NLM Scheme

Sl.	References, (Year)	Highlights of research work, remarks
	Conventional NLM Scheme	
1	[36], (2018)	A 13-level K-Type MLI with two DC sources to create a staircase waveform
2	[37], (2018)	A new Square T-type 17-level MLI topology
3	[38], (2018)	A three-phase MLI topology with a focus on reducing number of switches
4	[8], (2018)	A 49-level stacked MLI for drives application
	Research on improved/modified NLM Scheme	
5	[39], (2016)	Focus on providing a precise modulation index. But THD remains the same. No reduction in THD
6	[40], (2015)	Improved scheme to suit modular MLIs with upper and lower submodule addition. No clear explanation of extending its implementation on other MLI topologies
7	[41], (2015)	Focus of NLM on capacitor balancing in MLI. No generalized scheme applicable for all MLI topologies
	Implementation of NLM to Renewable Energy System	
8	[42], (2019)	Focus on asymmetrical CHB-MLI with 27-level output with PV input
9	[43], (2017)	Focus on modified CHB-MLI topology with PV source as input
10	[44], (2017)	Focus on a modified CCS-MLI topology with 19-level output

2.5 Concept of Modified Nearest Level Modulation Scheme (mNLM)

The current research trend in MLIs follows the conventional modulation schemes, and most of the research in modulation schemes have been carried out on high switching frequency. Hence, there is a desperate need for an improved conventional modulation scheme which operates under a low switching frequency. Some of the named parameters of improvement are reduction in THD, lower switching losses, ease and simplicity of the algorithm, a generalized solution to suit all the MLI topologies.

The modified Nearest Level Modulation Scheme (mNLM) takes the switching angles obtained by the conventional modulation scheme and proceeds by optimizing the switching angles. Theoretical calculation of the THD can be performed for a low switching frequency MLI output using Eqn. (2.4) [45]

$$THD = \frac{\sqrt{\frac{\pi^2 n^2}{8} - \frac{\pi}{4}\sum_{i=0}^{n-1}(2i+1)\alpha_{i+1} - (\sum_{i=1}^{n}cos(\alpha_i))^2}}{\sum_{i=1}^{n}cos(\alpha_i)} \tag{2.4}$$

The flowchart of mNLM algorithm shown in Fig. 2.3 is explained as follows [35]:

1. The mNLM process begins with initializing the number of levels (n) and the number of angles (i).

2. The next step starts with the iteration process with predefined values of switching angles. The values of the switching angles are calculated using Eqn. (2.3). This is the conventional method of NLM.

3. Once the THD for the initial set of switching angles is obtained, the iteration process begins by decreasing the first switching angle (α_1) by 1°. The other switching angles (α_2, α_3,...,α_n) are kept unchanged.

4. The revised THD is calculated with this newer set of switching angles. If the revised THD is found to be less than the previous one, the first switching angle (α_1) is further reduced by 1°.

5. In the second iteration, the revised THD is compared with the previous THD. This iteration process continues until there is no further reduction in the THD content when compared to the previous one.

6. If the revised THD is higher than the previous THD during the first iteration, (α_1) is kept the same and α_2 has proceeded with a decrease of 1°.

7. With the least THD obtained by varying the first switching angle (α_1) alone, the second switching (α_2) is decreased by 1° till the

least THD is obtained. This process continues for all the switching angles. This constitutes the first cycle.

8. Once the last switching angle (α_n) is varied, and the least THD is obtained, the entire process starting from (α_1) is repeated for the same $1°$ variation. This forms the second cycle.

9. The cycles are repeated until the THD content obtained is the minimum.

10. To fine-tune the switching angle values, the steps 3 to 8 are repeated with an accuracy of $0.10°$ for the newer set of switching angles obtained.

The mNLM algorithm is explained with the example shown in Table 2.3 where an MLI output of seven level is taken up. The initial switching angle values calculated are $\alpha_1 = 9.60°$, $\alpha_2 = 30.00°$ and $\alpha_3 = 56.44°$. From iterations 2 to 18, the switching angles are decreased by $1°$ whereas from iteration 19 to 27, the switching angles are decreased by $0.10°$. It took exactly 27 iterations to settle at the final switching angles of $\alpha_1 = 8.60°$, $\alpha_2 = 27.60°$ and $\alpha_3 = 50.44°$ to arrive at the least possible THD of 11.5344 %.

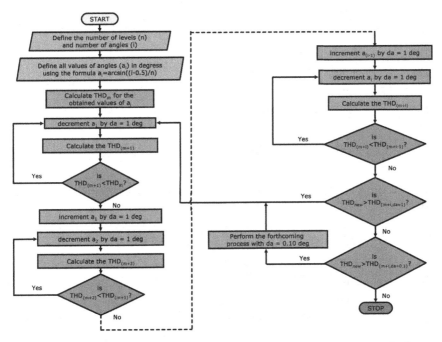

FIGURE 2.3: Algorithm of modified Nearest Level Algorithm [35]

TABLE 2.3: Complete Set of Iterations showing the Algorithm of mNLM for a 7-Level MLI Output

Iteration	α_1	α_2	α_3	THD	Remarks
1	9.60	30.00	56.44	12.2270	NLM firing angles
2	8.60	30.00	56.44	12.2429	THD ⇑ (dec. α by 1.00°)
3	9.60	29.00	56.44	12.1633	THD⇓
4	9.60	28.00	56.44	12.1816	THD⇑
5	9.60	29.00	55.44	11.9879	THD⇓
6	9.60	29.00	54.44	11.8447	THD⇓
7	9.60	29.00	53.44	11.7362	THD⇓
8	9.60	29.00	52.44	11.6647	THD⇓
9	9.60	29.00	51.44	11.6322	THD⇓
10	9.60	29.00	50.44	11.6402	THD⇑
11	8.60	29.00	51.44	11.6184	THD⇓
12	7.60	29.00	51.44	11.7088	THD⇑
13	8.60	28.00	51.44	11.5543	THD⇓
14	8.60	27.00	51.44	11.5755	THD⇑
15	8.60	28.00	50.44	11.5422	THD⇓
16	7.60	28.00	50.44	11.6243	THD⇑
17	8.60	27.00	50.44	11.5481	THD⇑
18	8.60	28.00	49.44	11.5725	THD⇑
19	8.50	28.00	50.44	11.5458	THD⇑ (dec. α by 0.10°)
20	8.60	27.90	50.44	11.5390	THD⇓
21	8.60	27.80	50.44	11.5366	THD⇓
22	8.60	27.70	50.44	11.5350	THD⇓
23	8.60	27.60	50.44	11.5344	Last significant iteration
24	8.60	27.50	50.44	11.5345	THD⇑
25	8.60	27.60	50.34	11.5349	THD⇑
26	8.50	27.60	50.44	11.5378	THD⇑
27	**8.60**	**27.60**	**50.44**	**11.5344**	**Final Values**

3

Implementation of LSF Modulation Schemes on Cross-Connected Sources based MLI

3.1 Cross-Connected Sources based MLI Topology

A single-phase Cross-Connected Sources based Multilevel Inverter (CCS-MLI) topology taken up, which is shown in Fig. 3.1. The CCS-MLI topology is composed of several isolated input DC sources. These DC sources are connected in a criss-cross manner with the help of the required number of power switches.

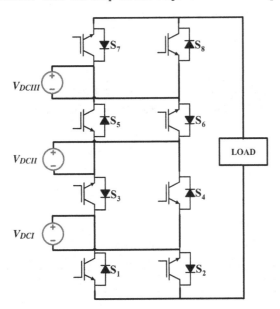

FIGURE 3.1: Single Phase Cross-Connected Sources based MLI Topology

The cross-connection in CCS-MLI topology is such that the positive terminal of a DC source is connected to the negative terminal of the next DC

19

source [46, 47]. This connection is repeated to successive and preceding DC
sources and hence, the name CCS-MLI topology.

The CCS-MLI topology presented in Fig. 3.1 consists of eight switches and
three DC sources. The topology can be operated in both symmetric as well
as asymmetric modes. There are four configurations with each successive configuration having the ability to generate AC output levels of a higher number
compared to the preceding configuration. The four configurations generate an
MLI level of 7-, 9-, 11- and 13-level. The modes of operation of each configuration are explained in the subsequent theory sections.

3.2 Working Principle of CCS-MLI Topology

The various configurations of CCS-MLI are tabulated in Table 3.1, which
shows the different possible switching configurations for generating each level.
For each configuration, the ratio of input DC sources are different, and hence,
the number of levels generated are different. There are four configurations
possible with the CCS-MLI topology. The symmetric and asymmetric configurations are explained in detail in the next sub-sections.

TABLE 3.1: Symmetric and Asymmetric Configurations of Operating CCS-
MLI Topology

Sl.	No. of Levels	Ratio of input DC sources	Possible Input DC Source combinations
1	7	1:1:1	V_{DCI}, $V_{DCI}+V_{DCII}$, $V_{DCI}+V_{DCII}+V_{DCIII}$
2	9	1:1:2	V_{DCI}, $V_{DCI}+V_{DCII}$, $V_{DCII}+V_{DCIII}$, $V_{DCI}+V_{DCII}+V_{DCIII}$
3	11	1:2:2	V_{DCI}, V_{DCII}, $V_{DCI}+V_{DCII}$, $V_{DCII}+V_{DCIII}$, $V_{DCI}+V_{DCII}+V_{DCIII}$
4	13	1:3:2	V_{DCI}, V_{DCIII}, V_{DCII}, $V_{DCI}+V_{DCII}$, $V_{DCII}+V_{DCIII}$, $V_{DCI}+V_{DCII}+V_{DCIII}$

3.2.1 Symmetric Configuration

The symmetric configuration carries all the input DC sources with an equal
voltage magnitude (V_{DCI}, V_{DCII}, V_{DCIII}). The peak of the output AC voltage is specified at 165 V, which allows for equal distribution of input DC

sources into three DC sources of 55 V each. The feature of symmetric configuration is that the switches distribute among themselves the voltage and current evenly. To add the various input DC sources to generate positive output, the switches $_2$, S_3, S_6 and S_7 are turned on. Similarly, negative output voltage is generated by turning on the switches S_1, S_4, S_5 and S_8. A 7-level output is obtained with the proposed symmetric configuration. The symmetric configuration can be extended to any number of levels with the addition of input DC source and power switches with the help of following equations.

The voltage magnitude of each input DC source ($V_{DC,j}$) is designed using Eqn. (3.1),

$$V_{DC,j} = V_{DC} \ for \ j = I, II, III, ..., n \tag{3.1}$$

where, j is the number of input DC voltage source.

Using Eqn. (3.2), the total number of levels (N_{level}) can be obtained for the symmetric configuration,

$$N_{level} = 2N_{DC} + 1 \tag{3.2}$$

where, N_{DC} is the total number of input DC voltage sources.

Using Eqn. (3.3), the required number of switches (N_{switch}) for a symmetric configuration is obtained,

$$N_{switch} = 2(N_{DC} + 1) \tag{3.3}$$

Using Eqn. (3.4), the maximum voltage V_{max} can be obtained which is the total of all the input DC voltage sources,

$$V_{max} = V_1 + V_2 + \cdots + V_j \tag{3.4}$$

The above mathematical expressions are useful to extend the CCS-MLI topology to any higher number of levels with a higher component count.

3.2.2 Asymmetric Configurations

In the asymmetric configuration, the values of input DC sources are unequal in magnitude. The asymmetric configuration generates levels of a higher number than the symmetric configuration. This enables the improvement in inverter efficiency to a great extent. With an increase in the number of levels generated by the MLI, there is a significant decrease in the Total Harmonics Distortion (THD) content.

In the first asymmetric configuration, the DC sources carry a fixed voltage magnitude ratio of 1:1:2. The value of the first and the second DC sources are 41.25 V while the third DC voltage source is rated at 82.5 V. Therefore, the third DC source is twice in magnitude compared to the other two DC sources. This configuration generates a 9-level output.

In the second asymmetric configuration, the voltage magnitude ratios of the DC sources are kept at 1:2:2. The first DC source is rated at 33 V. The other two input DC sources are rated at 66 V. Hence, the first DC source is half the other two DC sources in magnitude. An 11-level output is generated using this configuration.

In the third asymmetric configuration, a 1:3:2 voltage magnitude ratio of DC sources are considered. This resembles a natural number sequence in a slightly different arrangement. The DC sources carry voltages of different magnitudes. The first DC source is rated at 27.5 V, the second at 82.5 V and the third is rated at 55 V. This configuration generates a 13-level output. This is the maximum level that can be generated with the topology presented. Fig. 3.2 shows the complete specification to carry out simulation and experiment.

TABLE 3.2: Simulation and Experimental Specifications of CCS-MLI Topology

Sl.	Parameters	Specifications
1	Voltage/Current/Power from all DC sources	55 V/1.5 A/83 W
2	Output RMS Voltage	110 V
3	Output RMS Current	3.3 A
4	Output Power	360 W
5	Output AC frequency	50 Hz
6	Simulation Software	MATLAB/Simulink 2015b
7	DC sources	Aplab (64 V/5 A, 3 nos)
8	IGBT	H15R1203 (1200 V/30 A, 8 nos)
9	Opto-isolator	TLP250 (8 nos)
10	Controller	AVR Atmega32 Microcontroller
11	Loading Rheostat	35 ohms/5 A
12	Inductive Load	50 mH/5 A
13	Power Quality Analyzer	Fluke 435b

3.3   Simulation Results

This section presents simulation results of the Equal-Phase Method (EPM), Half-Equal Phase Method (HEPM), Selective Harmonics Elimination (SHE), Nearest Level Modulation (NLM) and modified Nearest Level Modulation (mNLM) schemes on CCS-MLI topology.

3.3.1 Single Phase Symmetric Configuration

Fig. 3.2 shows the 7-level single-phase simulated output voltage waveforms and the current waveforms for the symmetric configuration of CCS-MLI topology using R load for all the five modulation schemes. Small differences in switching angles of mNLM, NLM and SHE make the waveforms identical. But HEPM and EPM schemes have more significant differences in the switching angles, and hence, their waveform looks smaller in size. Fig. 3.3 shows the Fast Fourier Transform (FFT) analysis (fundamental voltage and voltage THD) of the output voltage for all the five modulation schemes.

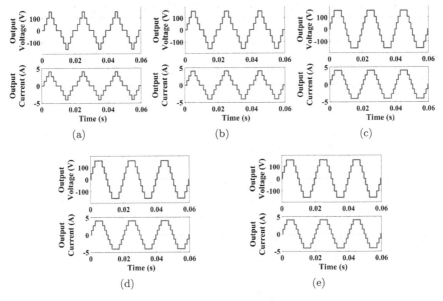

FIGURE 3.2: Simulated Output Voltage and Current Waveforms of Single Phase 7-Level CCS-MLI with R Load using (a) EPM, (b) HEPM, (c) SHE, (d) NLM and (e) mNLM

The simulation results are presented in the form of Table 3.3, which shows voltage THD, current THD, RMS voltage, RMS current, distortion factor and power as the inverter parameters for comparison. It can be seen that in all the inverter parameters, the mNLM scheme has performed better than the other modulation schemes. The voltage THD (10.45 %) and current THD (10.45 %) of mNLM are least among all the modulation schemes. The RMS voltage (117.2 V), RMS current (3.08 A) and output power (357 W) are also the highest among all the other modulation schemes.

Fig. 3.4 shows the 7-level single-phase simulated output voltage and current waveforms for all the five modulation schemes for the symmetric configuration of CCS-MLI topology using RL load. Fig. 3.5 and Fig. 3.6 presents the

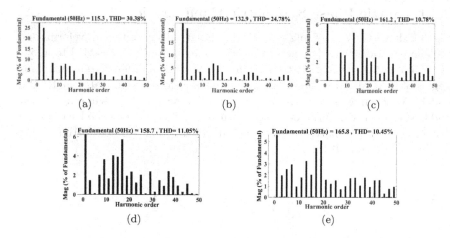

FIGURE 3.3: Simulated Output Voltage THD of Single Phase 7-Level CCS-MLI with R Load using (a) EPM, (b) HEPM, (c) SHE, (d) NLM and (e) mNLM

TABLE 3.3: Performance Comparison of Modulation Schemes with R Load on Single Phase CCS-MLI Topology—Simulation

Inverter Parameters	EPM	HEPM	SHE	NLM	mNLM
Voltage THD (%)	30.38	24.78	10.78	11.05	**10.45**
Current THD (%)	30.38	24.78	10.78	11.05	**10.45**
RMS Voltage (V)	81.5	94.0	114.0	112.2	**117.2**
RMS Current (A)	2.14	2.47	3.00	2.95	**3.08**
Distortion Factor	0.9155	0.9421	0.9885	0.9879	**0.9891**
Output Power (W)	160	219	338	327	**357**

FFT analysis of the output voltage and output current for all the five modulation schemes. The FFT analysis shows the fundamental voltage, voltage THD and current THD using each modulation scheme.

The simulation results are tabulated in Table 3.4, which shows voltage THD, current THD, RMS voltage, RMS current, power factor and power as the inverter parameters for comparison. The mNLM scheme has performed better than the other modulation schemes. The voltage THD (10.49 %) and current THD (1.76 %) of mNLM are least among all the modulation schemes. The RMS voltage (117.2 V) and RMS current (3.30 A) are also the highest among all the modulation schemes. The output power is found to be 352 W which is the highest among the modulation schemes. The next subsection deals with the various asymmetrical configurations of CCS-MLI topology and the associated simulation results and comparison of the performance analysis of mNLM with other modulation schemes.

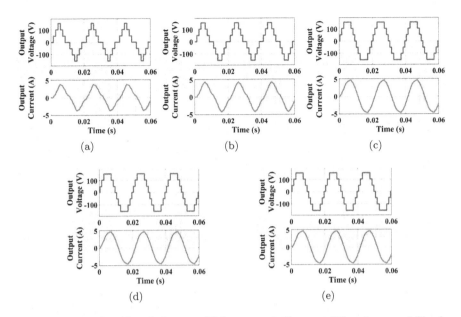

FIGURE 3.4: Simulated Output Voltage and Current Waveforms of Single Phase 7-Level CCS-MLI with RL Load using (a) EPM, (b) HEPM, (c) SHE, (d) NLM and (e) mNLM

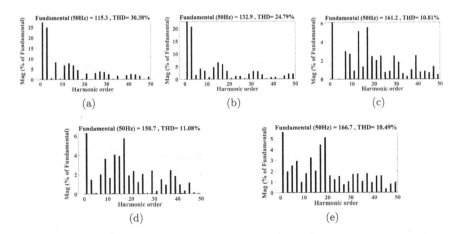

FIGURE 3.5: Simulated Output Voltage THD of Single Phase 7-Level CCS-MLI with RL Load using (a) EPM, (b) HEPM, (c) SHE, (d) NLM and (e) mNLM

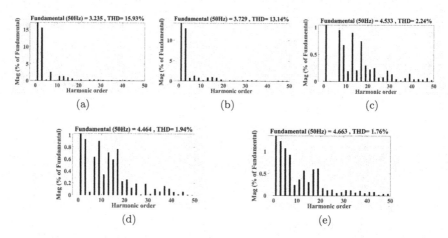

FIGURE 3.6: Simulated Output Current THD of Single Phase 7-Level CCS-MLI with RL Load using (a) EPM, (b) HEPM, (c) SHE, (d) NLM and (e) mNLM

TABLE 3.4: Performance Comparison of Modulation Schemes with RL Load on Single phase CCS-MLI Topology—Simulation

Inverter Parameters	EPM	HEPM	SHE	NLM	mNLM
Voltage THD (%)	30.38	24.79	10.81	11.08	**10.49**
Current THD (%)	15.93	13.14	2.24	1.94	**1.76**
RMS Voltage (V)	81.5	94.0	114.0	112.2	**117.2**
RMS Current (A)	2.29	2.64	3.21	3.16	**3.30**
Power Factor	0.8641	0.8800	0.9089	0.9087	**0.9093**
Output Power (W)	161	218	333	322	**352**

3.3.2 Single Phase Asymmetric Configurations

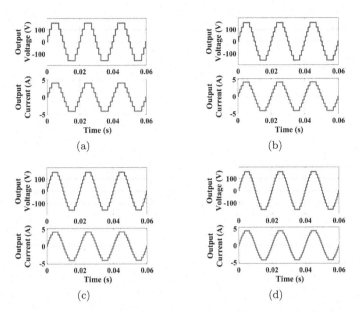

FIGURE 3.7: Simulated Output Voltage and Current Waveforms of Single Phase CCS-MLI with R Load using mNLM for (a) 7-Level, (b) 9-Level, (c) 11-Level and (d) 13-Level

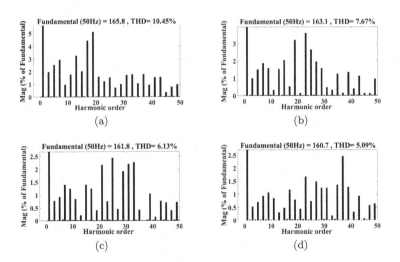

FIGURE 3.8: Simulated Output Voltage THD of Single Phase CCS-MLI Configurations with R Load using mNLM for (a) 7-Level, (b) 9-Level, (c) 11-Level and (d) 13-Level

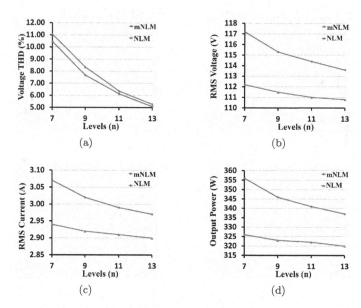

FIGURE 3.9: Simulated Parameters Comparison between mNLM and NLM (a) Voltage THD, (b) RMS Voltage, (c) RMS Current and (d) Output Power

Fig. 3.7 presents the simulation output using mNLM for the 7-, 9-, 11- and 13-level configuration. The MLI output approaches more and more of a sinewave shape as evident from the increase in the number of levels. Fig. 3.8 shows the FFT analysis of the four configurations. The fundamental voltage was found to be 165.8 V, 163.1 V, 161.8 V and 160.7 V for the 7-, 9-, 11- and 13-level configuration respectively. The voltage THD values were found to be 10.45%, 7.67%, 6.13% and 5.09% for the four configurations. Fig. 3.9 shows a comparison of simulation results between mNLM and NLM for four different inverter parameters, namely, voltage THD, RMS voltage, RMS current and output power. The following inferences can be concluded:

1. A healthy difference in the voltage THD between mNLM and NLM is seen. All configurations has a minimum difference of 0.25 %.

2. The RMS voltage is found to have increased with the use of mNLM compared to NLM scheme. An average increase of 5 V_{rms} was seen.

3. The RMS current is also seen as having a good increase in the magnitude of mNLM compared to NLM.

4. A good increase in the RMS voltage and RMS current has reflected in a significant rise in the output power by the use of the mNLM over NLM.

3.3.3 Three Phase Symmetric Configuration

The simulation of single-phase symmetric configuration of CCS-MLI topology is extended to three-phase star-connected configuration. The voltage and current output for the R and RL load are shown in Fig. 3.10. The simulation results are tabulated in Tables 3.5 and 3.6 for R and RL load respectively, where the proposed scheme is found to be effective among all the other modulation schemes.

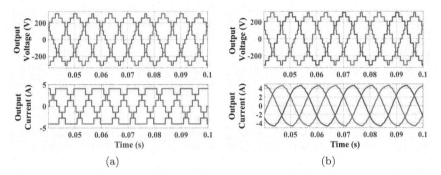

(a)   (b)

FIGURE 3.10: 3ϕ V & I Output Waveforms using mNLM with (a) R Load and (b) RL load

TABLE 3.5: Performance Comparison of Modulation Schemes with R Load on Three Phase CCS-MLI Topology—Simulation

Inverter Parameters	EPM	HEPM	SHE	NLM	mNLM
Voltage THD (%)	15.99	11.40	9.02	9.20	**8.27**
Current THD (%)	30.60	24.89	10.99	11.25	**10.72**
RMS Voltage (V)	143.9	165.5	201.0	198.0	**206.8**
RMS Current (A)	2.14	2.46	2.98	2.94	**3.07**
Distortion Factor	0.9442	0.9641	0.9899	0.9895	**0.9909**
Output Power (W)	291	393	593	576	**629**

TABLE 3.6: Performance Comparison of Modulation Schemes with RL Load on Three Phase CCS-MLI Topology—Simulation

Inverter Parameters	EPM	HEPM	SHE	NLM	mNLM
Voltage THD (%)	15.91	11.24	8.88	9.09	**8.11**
Current THD (%)	15.92	13.13	1.92	2.22	**1.72**
RMS Voltage (V)	143.9	165.5	201.0	197.9	**206.8**
RMS Current (A)	2.28	2.63	3.19	3.14	**3.28**
Power Factor	0.8919	0.9010	0.9107	0.9105	**0.9113**
Output Power (W)	293	392	584	566	**618**

3.4   Experimental Results

The experimental setup consists of three DC sources, the main circuit, an AVR programmer, an Atmega32 microcontroller, a load and a Fluke 435 power quality analyzer. The DC sources are variable in nature, and hence, for all the configurations, the same DC source with different voltage was used. The main circuit was fabricated on a printed circuit board. The TLP250 IC was used as the driver circuit to provide gate pulses to the switches. The AVR microcontroller has four ports with eight pins in each port. For an MLI, it is necessary to employ a microcontroller with a high number of output pins. Hence, the Atmega32 microcontroller was chosen. The Fluke 435b power quality analyzer measures the AC output voltage and current. It provides the precise THD of the AC waveform for 50 cycles.

This section presents the experimental results of Equal-Phase Method (EPM), Half-Equal Phase Method (HEPM), Selective Harmonics Elimination (SHE), Nearest Level Modulation (NLM) and modified Nearest Level Modulation (mNLM) schemes on CCS-MLI topology.

3.4.1   Single Phase Symmetric Configuration

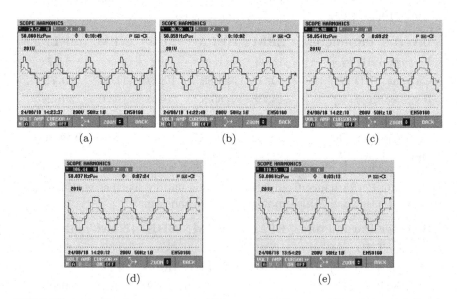

FIGURE 3.11: Hardware Output Voltage and Current Waveforms of Single Phase 7-Level CCS-MLI with R Load using (a) EPM, (b) HEPM, (c) SHE, (d) NLM and (e) mNLM

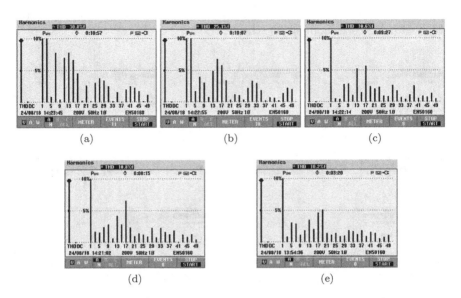

FIGURE 3.12: Voltage THD of Hardware Output of Single Phase 7-Level CCS-MLI with R Load using (a) EPM, (b) HEPM, (c) SHE, (d) NLM and (e) mNLM

TABLE 3.7: Performance Comparison of Modulation Schemes with R Load on Single Phase CCS-MLI Topology - Hardware

Inverter Parameters	EPM	HEPM	SHE	NLM	mNLM
Voltage THD (%)	30.8	25.1	10.6	10.8	**10.2**
Current THD (%)	31.0	25.3	10.7	10.8	**10.2**
RMS Voltage (V)	79.52	90.39	106.98	106.44	**110.35**
RMS Current (A)	2.4	2.7	3.2	3.2	**3.3**
Distortion Factor	0.9128	0.9402	0.9887	0.9884	**0.9897**
Output Power (W)	174	229	338	337	**360**

The Fig. 3.11 shows the 7-level single-phase hardware output voltage and current waveforms for all the five modulation schemes of the symmetric configuration of CCS-MLI topology using R load. It can be observed that the RMS voltage output has improved with the successive modulation scheme. The EPM and HEPM yielded a low RMS voltage of 79.52 V and 90.39 V, respectively. The SHE, NLM and mNLM schemes yielded a better RMS voltage close to 110 V. However, the proposed modulation scheme has yielded the highest RMS voltage output. A similar trend was observed with RMS current output. The mNLM scheme has yielded the maximum current of 3.3 A. The Fig. 3.12 shows the FFT analysis of the output voltage for all the five modulation schemes. The FFT analysis shows the fundamental voltage and voltage THD obtained using each modulation scheme.

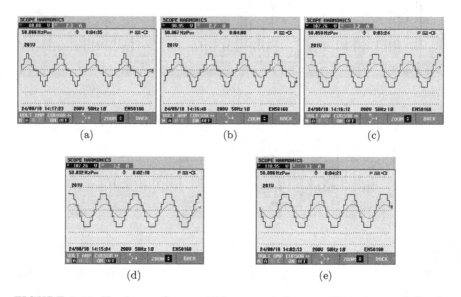

FIGURE 3.13: Hardware Output Voltage and Current Waveforms of Single Phase 7-Level CCS-MLI with RL Load using (a) EPM, (b) HEPM, (c) SHE, (d) NLM and (e) mNLM

The experimental results are tabulated in Table 3.7 where the voltage THD, current THD, RMS voltage, RMS current, distortion factor and power are compared. It can be seen that in all the inverter parameters, the mNLM scheme has performed better than the other modulation schemes. The voltage THD (10.2 %) and current THD (10.2 %) of mNLM are least among all the modulation schemes. The RMS voltage (110.35 V) and RMS current (3.3 A) is also the highest among all the modulation schemes. Since the RMS voltage and RMS current using mNLM are the highest, the output power (360 W) is also the highest among all the modulation schemes.

Fig. 3.13 shows the 7-level single-phase experimental output voltage and current waveforms for all the five modulation schemes for the symmetric configuration of CCS-MLI topology using RL load. Figs. 3.14 and 3.15 show the FFT analysis of the output voltage and output current for all the five modulation schemes. The FFT analysis shows the voltage THD and current THD using each modulation scheme.

The experimental results are tabulated in Table 3.8 where the voltage THD, current THD, RMS voltage, RMS current, power factor and power are the parameters considered for comparison. The performance of mNLM scheme is found to be better among all the modulation schemes. The voltage THD (10.6 %) and current THD (2.5 %) of mNLM are the least whiles, the RMS

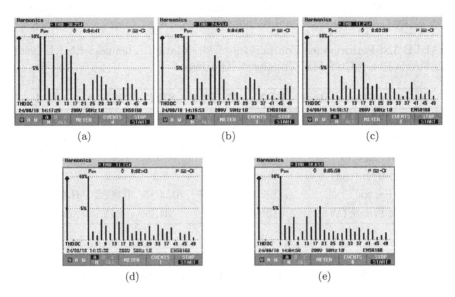

FIGURE 3.14: Voltage THD of Hardware Output of Single Phase 7-Level CCS-MLI with RL Load using (a) EPM, (b) HEPM, (c) SHE, (d) NLM and (e) mNLM

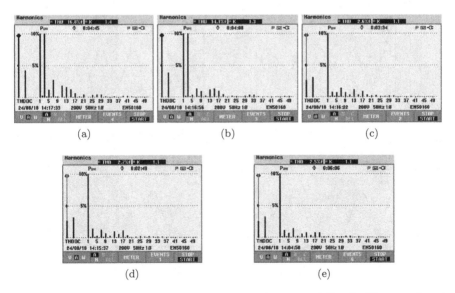

FIGURE 3.15: Current THD of Hardware Output of Single Phase 7-Level CCS-MLI with RL Load using (a) EPM, (b) HEPM, (c) SHE, (d) NLM and (e) mNLM

voltage (110.95 V), RMS current (3.30 A) and output power (333 W) are also the highest among all the modulation schemes.

TABLE 3.8: Performance Comparison of Modulation Schemes with RL Load on Single Phase CCS-MLI Topology—Hardware

Inverter Parameters	EPM	HEPM	SHE	NLM	mNLM
Voltage THD (%)	30.2	24.5	11.2	11.1	**10.6**
Current THD (%)	16.8	14.1	2.6	2.7	**2.5**
RMS Voltage (V)	80.08	90.95	107.76	107.26	**110.95**
RMS Current (A)	2.3	2.7	3.2	3.2	**3.3**
Power Factor	0.8633	0.8795	0.9085	0.9085	**0.9091**
Output Power (W)	159	216	313	312	**333**

3.4.2 Single Phase Asymmetric Configurations

Fig. 3.16 presents the hardware voltage and current waveforms output using mNLM for the 7-level, 9-level, 11-level and 13-level configuration. The RMS voltage was found to be 112.53 V, 110.90 V, 109.83 V and 108.91 V for the 7-, 9-, 11- and 13-level configuration respectively.

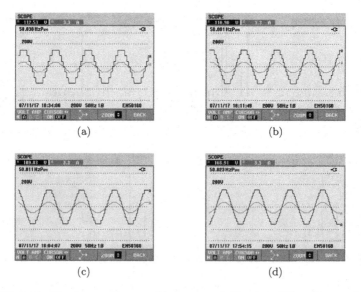

(a)   (b)

(c)   (d)

FIGURE 3.16: Hardware Output Voltage and Current Waveforms of Single Phase CCS-MLI Configurations with R Load using mNLM for (a) 7-Level, (b) 9-Level, (c) 11-Level and (d) 13-Level

Fig. 3.17 shows the FFT analysis of the four configurations. The voltage THD was found to be 10.2%, 7.5%, 6.1% and 5.2% for the four configurations. The comparison of experimental results between mNLM and NLM for four different inverter parameters, namely, voltage THD, RMS voltage, RMS current and output power are shown in Fig. 3.18. The following inferences can be concluded:

1. The mNLM scheme provides a decrease in the voltage THD by minimum 0.50% in all the configurations compared to NLM scheme.

2. A rise in the RMS voltage by 4 V_{rms} can be seen by the use of the mNLM scheme. This is an appreciable increase in the RMS voltage.

3. The proposed scheme also offers a noticeable rise in the RMS current of the AC output.

4. The output power can be seen with a significant rise due to the rise in RMS voltage and RMS current of AC output.

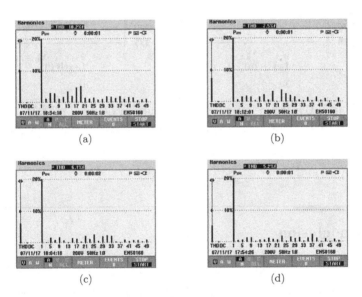

FIGURE 3.17: Hardware Output Voltage THD of Single Phase CCS-MLI Configurations with R Load using mNLM for (a) 7-Level, (b) 9-Level, (c) 11-Level and (d) 13-Level

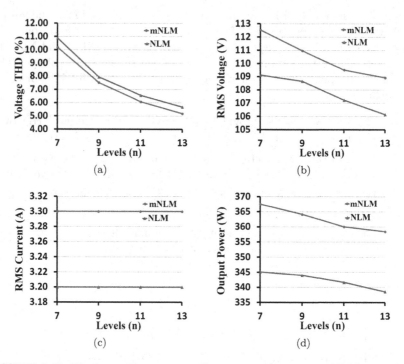

FIGURE 3.18: Hardware Parameters Comparison between mNLM and NLM
(a) Voltage THD, (b) RMS Voltage, (c) RMS Current and (d) Output Power

3.5   Conclusion

The implementation of LSF modulation schemes on CCS-MLI topology has
been presented. The NLM and mNLM scheme have been applied on the four
configurations of CCS-MLI topology for 7-, 9-, 11- and 13-level output. The
CCS-MLI topology was designed for a 360 W AC output system. The simulation was performed for symmetric and asymmetric configurations and extended to a three-phase system of symmetric configuration. The performance
was studied using a MATLAB/Simulink 2015b environment, and a hardware
prototype of the single-phase system of the same was demonstrated. The simulation and hardware experiments were carried out for R and RL load.

4

Implementation of LSF Modulation Schemes on Cascaded H-Bridge MLI

4.1   Cascaded H-Bridge MLI Topology

The equivalent circuit diagram of a CHBI is shown in Fig. 4.1. The CHBI presented in this chapter is composed of three input DC sources and twelve semiconductor switches. Three individual H-bridges are cascaded in series with each other to form CHBI.

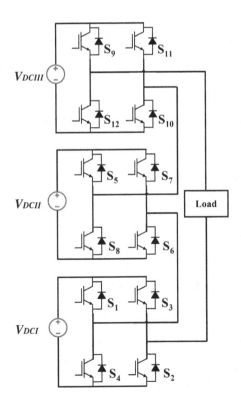

FIGURE 4.1: Cascaded H-bridge MLI Topology

The input DC voltage source from each H-bridge is reflected as a positive output at the load when switches S_1-S_2, S_5-S_6 and S_9-S_{10} are switched on. On the other hand, the same DC voltage source will be reflected as negative output when switches S_3-S_4, S_7-S_8 and S_{11}-S_{12} are switched on.

4.2 Working Principle of CHBI Topology

The various configurations of CHBI are tabulated in Table 4.1, which shows the different possible switching configurations for generating each level. The symmetric and asymmetric configurations are explained in detail in the next sub-sections.

4.2.1 Symmetric Configuration

When all the DC voltage sources are equal in magnitude, the result is a symmetric configuration of CHBI. Thus, a 7-level MLI output is obtained with three sources in a symmetric configuration. The three generated levels are V_{DCI}, V_{DCI}+V_{DCII} and V_{DCI}+V_{DCII}+V_{DCIII}. As the number of DC voltage sources and power switches is increased, it is possible to achieve a higher number of MLI levels.

The following set of generalized equations are useful to attain a higher number of the level using the symmetrical configuration [48].

The value of each DC voltage source is obtained by Eqn. (4.1),

$$V_{DC,j} = V_{DC} \ \ for \ \ j = I, II, III...\tag{4.1}$$

where j is the total count of input DC voltage sources.

Using Eqn. (4.2), the number of levels (N_{level}) can be obtained as,

$$N_{level} = 2x + 1 \ \ for \ \ x = 1, 2, 3...\tag{4.2}$$

where x is the H-bridge inverter numbers.

4.2.2 Asymmetric Configurations

Compared to a symmetric configuration, the switching pattern in an asymmetric configuration is slightly more complicated. The complexity of the switching pattern progressively increases with a higher number of levels.

The input DC sources in the first asymmetric configuration are in 1:1:2 voltage magnitude ratio. The third DC source is twice in magnitude to the first and second DC source. The first and second DC source is rated at 41.25 V while the third DC source is rated at 82.5 V. A 9-level output is generated using this configuration.

TABLE 4.1: Various Configurations of Operation of CHBI Circuit

Sl. No.	No. of Levels	Ratio of input DC sources	Possible Input DC Source combinations
1	7	1:1:1	$V_{DCI}, V_{DCI} + V_{DCII}, V_{DCI} + V_{DCII} + V_{DCIII}$
2	9	1:1:2	$V_{DCI}, V_{DCI} + V_{DCII}, V_{DCII} + V_{DCIII}$ $V_{DCI} + V_{DCII} + V_{DCIII}$
3	11	1:2:2	$V_{DCI}, V_{DCII}, V_{DCI} + V_{DCII}, V_{DCII} + V_{DCIII}$ $V_{DCI} + V_{DCII} + V_{DCIII}$
4	13	1:2:3	$V_{DCI}, V_{DCII}, V_{DCIII}, V_{DCI} + V_{DCIII},$ $V_{DCII} + V_{DCIII}, V_{DCI} + V_{DCII} + V_{DCIII}$
5	15	1:2:4	$V_{DCI}, V_{DCII}, V_{DCI} + V_{DCII}, V_{DCIII}, V_{DCI} + V_{DCIII},$ $V_{DCII} + V_{DCIII}, V_{DCI} + V_{DCII} + V_{DCIII}$
6	17	1:2:5	$V_{DCI}, V_{DCII}, V_{DCI} + V_{DCII}, V_{DCIII} - V_{DCI}, V_{DCIII},$ $V_{DCI} + V_{DCIII}, V_{DCII} + V_{DCIII}, V_{DCI} + V_{DCII} + V_{DCIII}$
7	19	1:2:6	$V_{DCI}, V_{DCII}, V_{DCI} + V_{DCII}, V_{DCIII} - V_{DCII},$ $V_{DCIII} - V_{DCI}, V_{DCIII}, V_{DCI} + V_{DCII},$ $V_{DCII} + V_{DCIII}, V_{DCI} + V_{DCII} + V_{DCIII}$
8	21	1:2:7	$V_{DCI}, V_{DCII}, V_{DCI} + V_{DCII}, V_{DCIII} - V_{DCII} - V_{DCI},$ $V_{DCIII} - V_{DCII}, V_{DCIII} - V_{DCI}, V_{DCIII}, V_{DCI} + V_{DCIII},$ $V_{DCII} + V_{DCIII}, V_{DCI} + V_{DCII} + V_{DCIII}$
9	23	1:3:7	$V_{DCI}, V_{DCII} - V_{DCI}, V_{DCIII}, V_{DCI} + V_{DCIII}, V_{DCI} +$ $V_{DCIII} - V_{DCII}, V_{DCIII} - V_{DCI}, V_{DCIII}, V_{DCI} + V_{DCIII},$ $V_{DCII} + V_{DCIII} - V_{DCI}, V_{DCII} + V_{DCIII}, V_{DCI} + V_{DCII} + V_{DCIII}$
10	25	1:3:8	$V_{DCI}, V_{DCII} - V_{DCI}, V_{DCIII}, V_{DCI} + V_{DCIII}, V_{DCIII} - V_{DCII},$ $V_{DCIII} + V_{DCI} - V_{DCII}, V_{DCIII} - V_{DCI}, V_{DCIII}, V_{DCI} + V_{DCIII},$ $V_{DCII} + V_{DCIII}, V_{DCI} + V_{DCII} + V_{DCIII}$
11	27	1:3:9	$V_{DCI}, V_{DCII} - V_{DCI}, V_{DCIII}, V_{DCI} + V_{DCII},$ $V_{DCIII} - V_{DCII} - V_{DCI}, V_{DCIII} - V_{DCII}, V_{DCIII} + V_{DCI} - V_{DCII},$ $V_{DCIII} - V_{DCI}, V_{DCIII}, V_{DCI} + V_{DCIII}, V_{DCII} + V_{DCIII},$ $V_{DCI} + V_{DCII} + V_{DCIII}$

The voltage magnitude ratios in the second asymmetric configuration are kept at 1:2:2 for the generation of an 11-level output. The first DC source is half in magnitude to the second and third DC source. The first DC source is rated at 33 V while the second and third DC source is rated at 66 V.

In the third asymmetric configuration, a voltage magnitude ratio of 1:2:3 is fixed for the input DC sources. This resembles a natural number sequence. The first input DC source is rated at 27.5 V, the second at 55 V and the third at 82.5 V. This configuration generates a 13-level output.

The fourth asymmetric configuration generates a 15-level output. The ratio of input DC sources is 1:2:4. The ratio resembles a binary state where the values of successive input DC sources are raised to the power of 2. The value of first, second and third DC sources are 24 V, 48 V and 96 V respectively. The fifth asymmetric configuration generates a 17-level output with 1:2:5 DC sources magnitude ratio. The value of each DC sources is 21 V, 42 V and 103 V. In the sixth asymmetric configuration, the input DC sources are arranged in 1:2:6 ratio to generate a 19-level MLI output. The input DC sources are 18 V, 36 V and 110 V.

An input DC sources ratio of 1:2:7 is used in the seventh configuration for the generation of a 21-level output. The input DC sources are rated at 16.5 V, 33 V and 115.5 V. In the eighth configuration, a 23-level MLI output is generated with a DC voltage ratio of 1:3:7. The values of the input DC sources are 15 V, 45 V and 105 V. The penultimate configuration generates a 25-level output carrying a DC voltage ratio of 1:3:8. The DC sources are rated at 14 V, 42 V and 110 V. In the tenth and last configuration, the DC sources are arranged in a trinary fashion of 1:3:9 for the generation of 27-level output with 13 V, 39 V and 115 V. All the configurations are simulated and validated experimentally using the proposed modulation scheme. The complete specification to carry out simulation and experimental results are shown in Table 4.2.

TABLE 4.2: Simulation and Experimental Specifications of CHBI Topology

Sl.	Parameters	Specifications
1	Voltage/Current/Power from all DC sources	55 V/1.6 A/90 W
2	Output RMS Voltage	110 V
3	Output RMS Current	3.5 A
4	Output Power	380 W
5	Output AC frequency	50 Hz
6	Simulation Software	MATLAB/Simulnk 2015b
7	DC sources	Aplab (64 V/5 A, 3 nos)
8	IGBT	H15R1203 (1200 V/30 A, 12 nos)
9	Opto-isolator	TLP250 (12 nos)
10	Controller	AVR Atmega32 Microcontroller
11	Loading Rheostat	32 ohms/5 A
12	Inductive Load	50 mH/5 A
13	Power Quality Analyzer	Fluke 435b

4.3   Simulation Results

The simulation results of all the five modulation schemes, namely, Equal-Phase Method (EPM), Half-Equal Phase Method (HEPM), Selective Harmonics Elimination (SHE), Nearest Level Modulation (NLM) and modified

Nearest Level Modulation (mNLM) schemes on CHBI topology are discussed in this section.

4.3.1   Single-phase Symmetric Configuration

Fig. 4.2 shows the 7-level single-phase simulated output voltage and current waveforms for all the five modulation schemes for the symmetric configuration of CHBI topology using the R load.

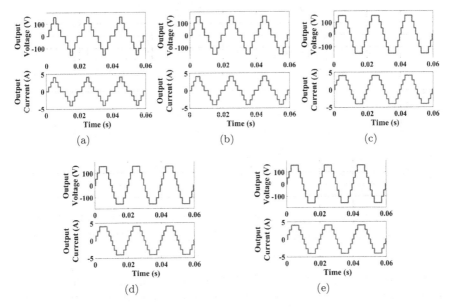

FIGURE 4.2: Simulated Output Voltage and Current Waveforms of Single-phase 7-Level CHBI with R Load using (a) EPM, (b) HEPM, (c) SHE, (d) NLM and (e) mNLM

TABLE 4.3: Performance Comparison of Modulation Schemes with R Load on Single-phase CHBI Topology—Simulation

Inverter Parameters	EPM	HEPM	SHE	NLM	mNLM
Voltage THD (%)	30.38	24.78	10.78	11.05	**10.45**
Current THD (%)	30.38	24.78	10.78	11.05	**10.45**
RMS Voltage (V)	81.5	94.0	114.0	112.2	**117.2**
RMS Current (A)	2.26	2.61	3.16	3.12	**3.26**
Distortion Factor	0.9155	0.9421	0.9885	0.9879	**0.9891**
Output Power (W)	169	231	356	346	**378**

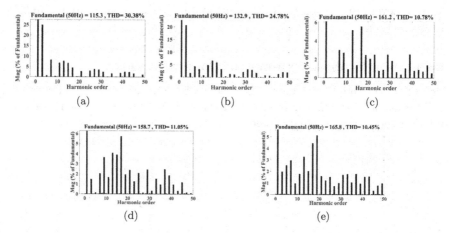

FIGURE 4.3: Simulated Output Voltage THD of Single-phase 7-Level CHBI with R Load using (a) EPM, (b) HEPM, (c) SHE, (d) NLM and (e) mNLM

The voltage and current waveforms of mNLM, NLM and SHE look alike due to the small differences in the switching angles while the HEPM and EPM schemes look distinct in shape due to substantial differences in the switching angles. Fig. 4.3 shows the Fast Fourier Transform (FFT) analysis (fundamental voltage and voltage THD) of the output voltage for all the five modulation schemes.

The simulation results are presented in Table 4.3, which shows the voltage THD, current THD, RMS voltage, RMS current, distortion factor and output power are the inverter parameters considered for comparison. It can be seen that, in all the inverter parameters, the mNLM scheme performed better than the other modulation schemes. The voltage THD (10.45 %) and current THD (10.45 %) of mNLM are the least among all the schemes. The RMS voltage (117.2 V) and RMS current (3.26 A) is also the highest among all the other modulation schemes. The output power (378 W) is also the highest among all the other modulation schemes.

Fig. 4.4 shows the 7-level single-phase simulated output voltage and current waveforms for all the five modulation schemes for the symmetric configuration of CHBI topology using RL load. Figs. 4.5 and 4.6 show the FFT analysis of the output voltage and output current for all the five modulation schemes. The FFT analysis shows the fundamental voltage, voltage THD and current THD using each modulation scheme.

Table 4.4 shows the simulation results of the voltage THD, current THD, RMS voltage, RMS current, power factor and power as the inverter parameters considered for comparison. It can be seen that, in all the inverter parameters, the mNLM scheme has performed better than the other modulation schemes. The voltage THD (10.51 %) and current THD (1.78 %) of mNLM are the

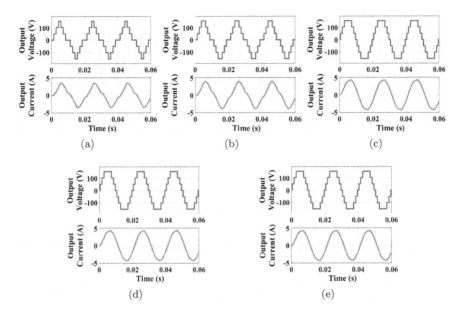

FIGURE 4.4: Simulated Output Voltage and Current Waveforms of Single-phase 7-Level CHBI with RL Load using (a) EPM, (b) HEPM, (c) SHE, (d) NLM and (e) mNLM

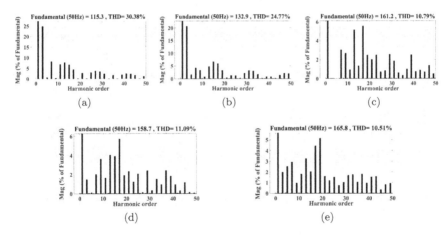

FIGURE 4.5: Simulated Output Voltage THD of Single-phase 7-Level CHBI with RL Load using (a) EPM, (b) HEPM, (c) SHE, (d) NLM and (e) mNLM

least among all the modulation schemes. The RMS voltage (117.2 V) and RMS current (3.06 A) are also the highest among all the other modulation schemes. The output power (321 W) is also the highest among all the other

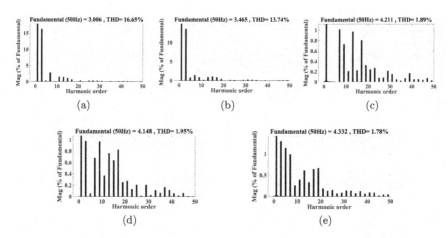

FIGURE 4.6: Simulated Output Current THD of Single-phase 7-Level CHBI with RL Load using (a) EPM, (b) HEPM, (c) SHE, (d) NLM and (e) mNLM

modulation schemes. The next subsection deals with the various asymmetrical configurations of CHBI topology and the associated simulation results and performance analysis comparison of mNLM with other modulation schemes.

TABLE 4.4: Performance Comparison of Modulation Schemes with RL Load on Single-phase CHBI Topology—Simulation

Inverter Parameters	EPM	HEPM	SHE	NLM	mNLM
Voltage THD (%)	30.38	24.77	10.79	11.09	**10.51**
Current THD (%)	16.65	13.74	1.89	1.95	**1.78**
RMS Voltage (V)	81.5	94.0	114.0	112.2	**117.3**
RMS Current (A)	2.12	2.45	2.98	2.93	**3.06**
Power Factor	0.8491	0.8654	0.8946	0.8943	**0.8949**
Output Power (W)	147	199	304	294	**321**

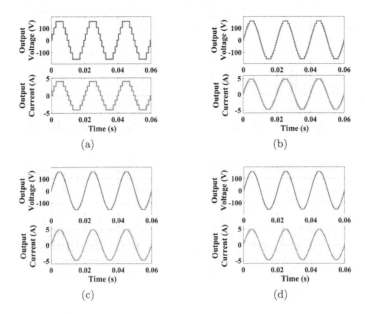

FIGURE 4.7: Simulated Output Voltage and Current Waveforms of Single-phase CHBI with R Load using mNLM for (a) 7-Level, (b) 15-Level, (c) 21-Level and (d) 27-Level

4.3.2 Single-phase Asymmetric Configurations

Fig. 4.7 presents the simulation output using mNLM for the 7-level, 15-level, 21-level and 27-level configuration. With the increase in the number of levels, it is observed that the MLI output approaches more and more of a sinewave shape.

Fig. 4.8 shows the FFT analysis of the four configurations. The fundamental voltage was found to be 165.8 V, 160.1 V, 158.8 V and 157.9 V for the 7-, 15-, 21- and 27-level configuration respectively. The voltage THD was found to be 10.45%, 4.34%, 2.32% and 1.45% for the four configurations, respectively.

Fig. 4.9 shows a comparison of simulated results between mNLM and NLM for five different inverter parameters, namely, voltage THD, RMS voltage, RMS current and output power. The following conclusions are drawn:

1. A decrease in THD by 0.20% is provided by the proposed algorithm over the NLM in all the configurations.

2. The mNLM scheme can increase the RMS voltage at the AC output by an average of 4 V_{RMS} over the conventional scheme.

3. The increase in the RMS current by the use of mNLM is also noticed.

4. At the end, the output power has found an increase due to the increase in RMS voltage and RMS current.

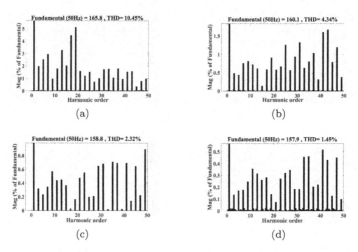

(a)

(b)

(c)

(d)

FIGURE 4.8: Simulated Output Voltage THD of Single-phase CHBI with R Load using mNLM for (a) 7-Level, (b) 15-Level, (c) 21-Level and (d) 27-Level

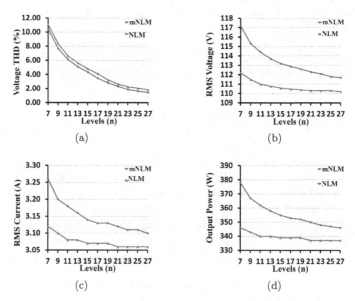

(a)

(b)

(c)

(d)

FIGURE 4.9: Simulated Parameters Comparison between mNLM and NLM (a) Voltage THD, (b) RMS Voltage, (c) RMS Current and (d) Output Power

4.3.3 Three Phase Symmetric Configuration

The simulation of single-phase symmetric configuration of CHBI topology is extended to three-phase star-connected configuration. The voltage and current output for the R and RL load are shown in Fig. 4.10. The simulation results are tabulated in Tables 4.5 and 4.6 for R and RL load respectively, where the proposed scheme is found to be effective among all the other modulation schemes.

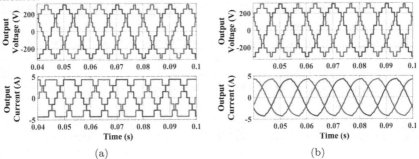

(a)   (b)

FIGURE 4.10: 3 ϕ V and I Output Waveforms using mNLM with (a) R Load and (b) RL Load

TABLE 4.5: Performance Comparison of Modulation Schemes with R Load on Three Phase CHBI Topology—Simulation

Inverter Parameters	EPM	HEPM	SHE	NLM	mNLM
Voltage THD (%)	16.00	11.41	9.01	9.28	**8.28**
Current THD (%)	30.39	24.96	11.01	11.24	**10.68**
RMS Voltage (V)	143.9	165.5	200.9	197.8	**206.7**
RMS Current (A)	2.26	2.60	3.15	3.11	**3.25**
Distortion Factor	0.9447	0.9639	0.9899	0.9894	**0.9909**
Output Power (W)	307	415	626	609	**666**

TABLE 4.6: Performance Comparison of Modulation Schemes with RL Load on Three Phase CHBI Topology—Simulation

Inverter Parameters	EPM	HEPM	SHE	NLM	mNLM
Voltage THD (%)	15.88	11.26	8.90	9.14	**8.18**
Current THD (%)	16.66	13.73	2.06	2.35	**1.86**
RMS Voltage (V)	143.8	165.6	200.9	197.9	**206.7**
RMS Current (A)	2.12	2.44	2.96	2.92	**3.05**
Power Factor	0.8767	0.8860	0.8962	0.8960	**0.8968**
Output Power (W)	267	358	533	518	**565**

4.4 Experimental Results

The hardware setup consists of three DC sources, three H-bridge cascaded with each other, an AVR programmer, Atmega32 microcontroller, a load and a Fluke 435 power quality analyzer. Similar to the previous topology, the DC sources are variable in nature. This helps in employing the same DC source for all the configurations. The primary circuit is a combination of three H-bridge circuit cascaded with each other. Eight pins of Port B of Atmega32 microcontroller are employed for the generation of the gate pulses. The gate pulses from microcontroller are fed to the input pins of the TLP250 driver circuit to operate the semiconductor switches. The Fluke 435b power quality analyzer helps in measuring the hardware voltage THD and current THD.

The experimental results of Equal-Phase Method (EPM), Half-Equal Phase Method (HEPM), Selective Harmonics Elimination (SHE), Nearest Level Modulation (NLM) and modified Nearest Level Modulation (mNLM) schemes on CHBI topology for symmetric and asymmetric configurations are presented in this section.

4.4.1 Single-phase Symmetric Configuration

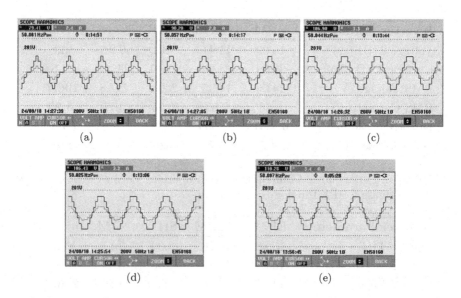

FIGURE 4.11: Hardware Output Voltage and Current Waveforms of Single-phase 7-Level CHBI with R Load using (a) EPM, (b) HEPM, (c) SHE, (d) NLM and (e) mNLM

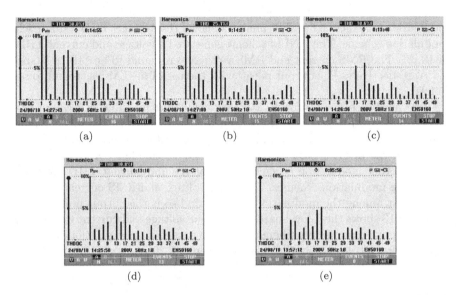

FIGURE 4.12: Voltage THD of Hardware Output of Single-phase 7-Level CHBI with R Load using (a) EPM, (b) HEPM, (c) SHE, (d) NLM and (e) mNLM

The Fig. 4.11 shows the 7-level hardware output voltage and current waveforms for all the five modulation schemes for the symmetric configuration of CHBI topology using R load. It can be observed from the output that the EPM and HEPM waveforms generated a low RMS voltage and RMS current. This is naturally because of the poor waveform, which looks similar to the triangular waveform. When the other three modulation schemes are observed, the RMS voltage and RMS current found to be good. This is due to approaching the sinusoidal nature of waveform. However, among the three modulation schemes, the proposed modulation scheme generated the highest RMS voltage and RMS current.

TABLE 4.7: Performance Comparison of Modulation Schemes with R Load on Single-phase CHBI Topology—Hardware

Inverter Parameters	EPM	HEPM	SHE	NLM	mNLM
Voltage THD (%)	30.8	25.1	10.6	10.8	**10.2**
Current THD (%)	30.9	25.3	10.7	10.9	**10.2**
RMS Voltage (V)	79.41	90.28	106.90	106.43	**110.28**
RMS Current (A)	2.4	2.8	3.3	3.3	**3.4**
Distortion Factor	0.9130	0.9402	0.9887	0.9883	**0.9897**
Output Power (W)	174	238	349	347	**371**

Fig. 4.12 shows the FFT analysis of the output voltage for all the five modulation schemes. The FFT analysis shows the voltage THD using each modulation scheme. The mNLM scheme generated a voltage and current THD of 10.2%. The RMS current and RMS voltage were found to be 110.28 V and 3.4 A, respectively. The output power with the use of mNLM was found to be 371 W.

The hardware results for R load are tabulated in Table 4.7 where the voltage THD, current THD, RMS voltage, RMS current, distortion factor and power are the inverter parameters considered for comparison. It is seen that in all the inverter parameters, the mNLM scheme has performed better than the other modulation schemes. The voltage and current THD are least among the modulation schemes. The RMS voltage and RMS current are also the highest among the modulation schemes and so is the output power.

Fig. 4.13 shows the 7-level hardware output voltage and current waveforms for all the five modulation schemes for the symmetric configuration of CHBI topology for RL load. The RL load was kept at 29 ohms and 50 mH. The difference in the shapes of the waveforms can be observed among the five modulation schemes, especially the HEPM and EPM schemes.

Figs. 4.14 and 4.15 shows the FFT analysis of the output voltage and output current for all the five modulation schemes. The FFT analysis shows the fundamental voltage, voltage THD and current THD. The mNLM scheme generated a voltage and current THD of 10.6% and 2.5% respectively. The

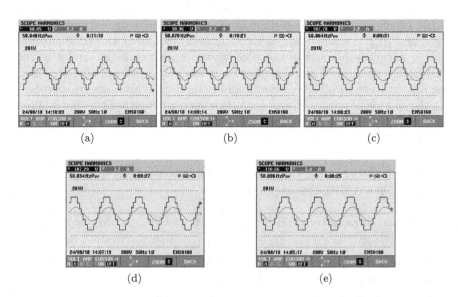

FIGURE 4.13: Hardware Output Voltage and Current Waveforms of Single-phase 7-Level CHBI with RL Load using (a) EPM, (b) HEPM, (c) SHE, (d) NLM and (e) mNLM

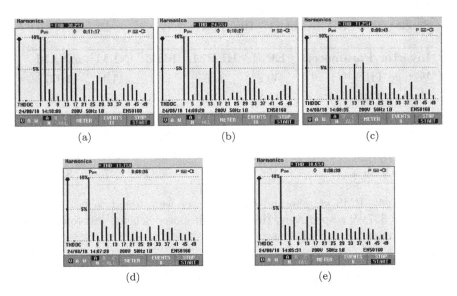

FIGURE 4.14: Voltage THD of Hardware Output of Single-phase 7-Level CHBI with RL Load using (a) EPM, (b) HEPM, (c) SHE, (d) NLM and (e) mNLM

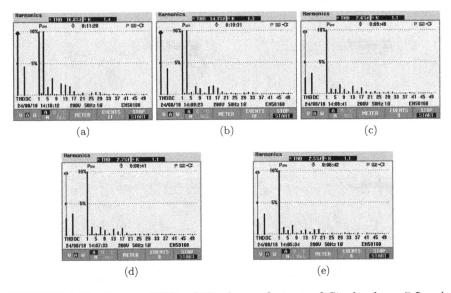

FIGURE 4.15: Current THD of Hardware Output of Single-phase 7-Level CHBI with RL Load using (a) EPM, (b) HEPM, (c) SHE, (d) NLM and (e) mNLM

TABLE 4.8: Performance Comparison of Modulation Schemes with RL Load on Single-phase CHBI Topology—Hardware

Inverter Parameters	EPM	HEPM	SHE	NLM	mNLM
Voltage THD (%)	30.2	24.5	11.2	11.1	**10.6**
Current THD (%)	16.8	14.1	2.6	2.7	**2.5**
RMS Voltage (V)	80.05	90.96	107.78	107.29	**110.66**
RMS Current (A)	2.3	2.7	3.2	3.2	**3.4**
Power Factor	0.8496	0.8655	0.8941	0.8941	**0.8947**
Output Power (W)	156	216	308	307	**337**

RMS current, RMS voltage and output power were found to be 110.66 V, 3.4 A and 337 W respectively. Hence, the proposed scheme was found to be effective among all the schemes.

4.4.2   Single-phase Asymmetric Configurations

Fig. 4.16 presents the hardware voltage and current waveforms output using mNLM for the 7-, 15-, 21- and 27-level configuration. The RMS voltage was found to be 112.33 V, 108.46 V, 107.72 V and 106.39 V for the 7-, 15-, 21- and 27-level configuration respectively.

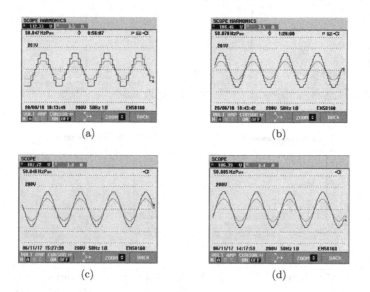

(a)

(b)

(c)

(d)

FIGURE 4.16: Hardware Output Voltage and Current Waveforms of Single-phase CHBI with R Load using mNLM for (a) 7-Level, (b) 15-Level, (c) 21-Level and (d) 27-Level

Fig. 4.17 shows the FFT analysis of the four configurations. The voltage THD were 10.1%, 3.9%, 2.5% and 2.2% for the four configurations. Fig. 4.18 shows a comparison of experimental results between mNLM and NLM for four different inverter parameters, namely, voltage THD, RMS voltage, RMS current and output power. The following conclusions have been drawn:

1. In all the configurations, the mNLM scheme offers a reduced voltage THD of minimum 0.50% for all the configurations.

2. An average increase of 4 V_{RMS} was observed in the RMS voltage for all the configurations.

3. The mNLM scheme also generates a higher RMS current than the NLM scheme.

4. The output power using mNLM was found to be sufficiently higher than the NLM scheme.

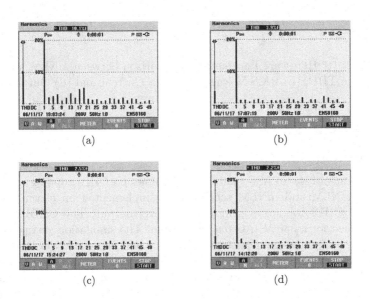

FIGURE 4.17: Voltage THD of Hardware Output of Single-phase CHBI with R Load using mNLM for (a) 7-Level, (b) 15-Level, (c) 21-Level and (d) 27-Level

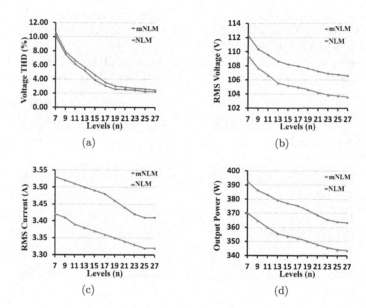

FIGURE 4.18: Hardware Parameters Comparison between mNLM and NLM (a) Voltage THD, (b) RMS Voltage, (c) RMS Current and (d) Output Power

4.5 Conclusion

The implementation of LSF modulation schemes on CHBI topology has been presented. NLM and mNLM scheme were implemented on eleven configurations of CHBI topology, from 7- to 27-level output. The CHBI topology was designed for a 380 W AC output system. The simulation was performed for symmetric and asymmetric configurations and extended to a three-phase system of symmetric configuration. The performance was studied using a MATLAB/Simulink 2015b environment, and a hardware prototype of the single-phase system of the same was demonstrated. The simulation and hardware experiments were carried out for R and RL load.

5

Implementation of LSF Modulation Schemes on Multilevel DC-Link Inverter

5.1 Multilevel DC-Link Inverter Topology

The Cascaded H-bridge (CHB) Inverter was operated from 7- to 27-level for eleven different configurations. This chapter presents a Multilevel DC-Link Inverter (MLDCLI) operating under twelve different configurations. The configurations presented start from a 9-level MLI configuration to a 31-level MLI configuration. Figure 5.1 shows the MLDCLI topology where twelve power switches along with four DC sources are used. An MLDCLI topology is basically a combination of several half-bridge circuits in series and a single H-bridge.

A half-bridge is a combination of two switches which operate in a toggle manner. Considering the switch pair S_1-S_2, the voltage source V_{DCI} can be added to the main circuit by turning on the switch S_2 and turning off switch S_1. For keep the voltage source V_{DCI} out of the main circuit, the switch S_2 is turned off and switch S_1 is turned on. Similarly, for all the half-bridge circuits, the association and disassociation of the voltage sources follow a similar manner of operation. The h-bridge designed with switches S_9, S_{10}, S_{11}, S_{12} provides the positive and negative polarity to generate a complete cycle of AC waveform.

5.2 Working Principle of MLDCLI

5.2.1 Symmetric Configuration

A symmetric MLI configuration utilizes various input DC sources of equal voltage magnitude. This leads to four possible voltage levels i.e, V_{DCI}, $V_{DCI}+V_{DCII}$, $V_{DCI}+V_{DCII}+V_{DCIII}$ and $V_{DCI}+V_{DCII}+V_{DCIII}+V_{DCIV}$. This first configuration generated a 9-level MLI output [49] have shown that this symmetrical configuration can be expanded with the following equations.

As all the DC sources are equal in magnitude, the value of the DC sources are represented by Eqn. (5.1).

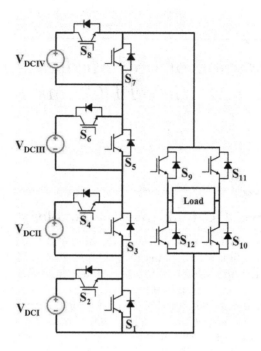

FIGURE 5.1: Multilevel DC-link Inverter MLI Topology

$$V_{DC,j} = V_{DC} \ \ for \ \ j = I, II, III, IV... \tag{5.1}$$

Where j is the number of DC voltage source.

The number of switches and the number of MLI levels in the MLDCLI is represented as Eqn. (5.2).

$$N_{level} = n - 3 \ \ for \ \ n = 6, 8, 10, 12.... \tag{5.2}$$

Where N_{level} is the number of levels, and n is the number of switches.

For achieving a higher number of MLI levels, the number of sources required is higher, and so is the number of input sources shown in Eqn. (5.3).

$$N_{source} = \frac{n - 4}{2} \ \ for \ \ n = 6, 8, 10, 12.... \tag{5.3}$$

Where N_{source} is the number of sources.

5.2.2 Asymmetric Configurations

The asymmetric configurations have input DC voltage sources of different voltage magnitude with each source being a multiple of first DC source, which

TABLE 5.1: Various Configurations of Operation of MLDCLI (1 to 7)

Sl. No. of	Levels	Ratio of DC sources	Possible Voltage combinations
1	9	1:1:1:1	V_{DCI}, $V_{DCI} + V_{DCII}$,
			$V_{DCI} + V_{DCII} + V_{DCIII}$,
			$V_{DCI} + V_{DCII} + V_{DCIII} + V_{DCIV}$
2	11	1:1:1:2	V_{DCI}, $V_{DCI} + V_{DCII}$,
			$V_{DCI} + V_{DCII} + V_{DCIII}$,
			$V_{DCII} + V_{DCIII} + V_{DCIV}$,
			$V_{DCI} + V_{DCII} + V_{DCIII} + V_{DCIV}$
3	13	1:1:2:2	V_{DCI}, $V_{DCI} + V_{DCII}$, $V_{DCII} + V_{DCIII}$,
			$V_{DCI} + V_{DCII} + V_{DCIII}$,
			$V_{DCII} + V_{DCIII} + V_{DCIV}$,
			$V_{DCI} + V_{DCII} + V_{DCIII} + V_{DCIV}$
4	15	1:2:2:2	V_{DCI}, V_{DCII}, $V_{DCI} + V_{DCII}$,
			$V_{DCII} + V_{DCIII}$, $V_{DCI} + V_{DCII} + V_{DCIII}$,
			$V_{DCII} + V_{DCIII} + V_{DCIV}$, $V_{DCI} + V_{DCII} + V_{DCIII} + V_{DCIV}$
5	17	1:2:2:3	V_{DCI}, V_{DCII}, $V_{DCI} + V_{DCII}$,
			$V_{DCII} + V_{DCIII}$, $V_{DCI} + V_{DCII} + V_{DCIII}$,
			$V_{DCI} + V_{DCII} + V_{DCIV}$, $V_{DCII} + V_{DCIII} + V_{DCIV}$,
			$V_{DCI} + V_{DCII} + V_{DCIII} + V_{DCIV}$
6	19	1:2:2:4	V_{DCI}, V_{DCII}, $V_{DCI} + V_{DCII}$,
			$V_{DCII} + V_{DCIII}$, $V_{DCI} + V_{DCII} + V_{DCIII}$,
			$V_{DCII} + V_{DCIV}$, $V_{DCI} + V_{DCII} + V_{DCIV}$,
			$V_{DCII} + V_{DCIII} + V_{DCIV}$, $V_{DCI} + V_{DCII} + V_{DCIII} + V_{DCIV}$
7	21	1:2:3:4	V_{DCI}, V_{DCII}, V_{DCIII}, V_{DCIV}, $V_{DCI} + V_{DCIV}$,
			$V_{DCII} + V_{DCIV}$, $V_{DCIII} + V_{DCIV}$, $V_{DCI} + V_{DCIII} + V_{DCIV}$,
			$V_{DCII} + V_{DCIII} + V_{DCIV}$, $V_{DCI} + V_{DCII} + V_{DCIII} + V_{DCIV}$

is dependent on the type of configuration. An MLDCLI topology can be operated with any asymmetrical configuration depending on the number of input DC sources and number of switches. For achieving a better quality AC output, it is preferred to go with a asymmetrical configuration which offers the highest number of MLI levels. In this section, there are 11 asymmetrical configurations considered for the generation of levels starting from 11 to 31. An insight into the modes of operation of each configuration is also presented. The proposed modulation scheme is implemented on all the symmetrical as well as asymmetrical configurations.

In configuration 2, the first three DC sources are equal in magnitude, but the fourth DC voltage source is twice the voltage magnitude resulting in an

TABLE 5.2: Various Configurations of Operation of MLDCLI (8 to 12)

Sl. No. of Levels	Ratio of DC sources	Possible Voltage combinations
8	23	1:2:3:5	V_{DCI}, V_{DCII}, V_{DCIII}, $V_{DCI} + V_{DCIII}$, V_{DCIV}, $V_{DCI} + V_{DCIV}$, $V_{DCII} + V_{DCIV}$, $V_{DCIII} + V_{DCIV}$, $V_{DCI} + V_{DCIII} + V_{DCIV}$, $V_{DCII} + V_{DCIII} + V_{DCIV}$, $V_{DCI} + V_{DCII} + V_{DCIII} + V_{DCIV}$
9	25	1:2:3:6	V_{DCI}, V_{DCII}, V_{DCIII}, $V_{DCI} + V_{DCIII}$, $V_{DCII} + V_{DCIII}$, $V_{DCI} + V_{DCII} + V_{DCIII}$, $V_{DCI} + V_{DCIV}$, $V_{DCII} + V_{DCIV}$, $V_{DCIII} + V_{DCIV}$, $V_{DCI} + V_{DCIII} + V_{DCIV}$, $V_{DCII} + V_{DCIII} + V_{DCIV}$, $V_{DCI} + V_{DCII} + V_{DCIII} + V_{DCIV}$
10	27	1:2:4:6	V_{DCI}, V_{DCII}, $V_{DCI} + V_{DCII}$, V_{DCIII}, $V_{DCI} + V_{DCIII}$, $V_{DCII} + V_{DCIII}$, $V_{DCI} + V_{DCII} + V_{DCIII}$, $V_{DCII} + V_{DCIV}$, $V_{DCI} + V_{DCII} + V_{DCIV}$, $V_{DCIII} + V_{DCIV}$, $V_{DCI} + V_{DCIII} + V_{DCIV}$, $V_{DCII} + V_{DCIII} + V_{DCIV}$, $V_{DCI} + V_{DCII} + V_{DCIII} + V_{DCIV}$
11	29	1:2:4:7	V_{DCI}, V_{DCII}, $V_{DCI} + V_{DCII}$, V_{DCIII}, $V_{DCI} + V_{DCIII}$, $V_{DCII} + V_{DCIII}$, $V_{DCI} + V_{DCII} + V_{DCIII}$, $V_{DCI} + V_{DCIV}$, $V_{DCII} + V_{DCIV}$, $V_{DCI} + V_{DCII} + V_{DCIV}$, $V_{DCIII} + V_{DCIV}$, $V_{DCI} + V_{DCIII} + V_{DCIV}$, $V_{DCII} + V_{DCIII} + V_{DCIV}$, $V_{DCI} + V_{DCII} + V_{DCIII} + V_{DCIV}$
12	31	1:2:4:8	V_{DCI}, V_{DCII}, $V_{DCI} + V_{DCII}$, V_{DCIII}, $V_{DCI} + V_{DCIII}$, $V_{DCII} + V_{DCIII}$, $V_{DCI} + V_{DCII} + V_{DCIII}$, V_{DCIV}, $V_{DCI} + V_{DCIV}$, $V_{DCII} + V_{DCIV}$, $V_{DCI} + V_{DCII} + V_{DCIV}$, $V_{DCIII} + V_{DCIV}$, $V_{DCI} + V_{DCIII} + V_{DCIV}$, $V_{DCII} + V_{DCIII} + V_{DCIV}$, $V_{DCI} + V_{DCII} + V_{DCIII} + V_{DCIV}$

11-level MLI output. The DC sources are rated at 33 V, 33 V, 33 V and 66 V. In the third configuration, the first two DC sources are equal in magnitude while the other two sources are double in voltage magnitude with input DC sources of 27.5 V, 27.5 V, 55 V and 55 V. This results in a 13-level output. For generate 15-level, the DC input voltage sources are in 1:2:2:2 ratio where each DC source is set at 24 V, 48 V, 48 V and 48 V. With an input DC voltage sources of 1:2:2:3 and 1:2:2:4, a 17- and 19-level MLI output is generated. For 17-level the DC sources are 21 V, 42 V, 42 V and 63 V while for the 19 levels, the DC sources are 18 V, 36 V, 36 V and 73 V. The DC voltage sources follow a natural number sequence for the generation of a 21-level MLI output. The DC voltages for this are 16.5 V, 33 V, 50 V and 66 V. The DC input voltage sources follows a Fibonacci series of 1:2:3:5 for the generation of 23-level MLI output. The value of DC input sources are 15 V, 30 V, 45 V and 75 V. With the input DC voltage sources ratio of 1:2:3:6 carrying 14 V, 28 V, 41 V and 82 V, 25-level MLI is generated. A 1:2:4:6 and 1:2:4:7 ratios of DC voltage sources generate a 27-level and 29-level MLI output, respectively. The DC sources are rated at 13 V, 26 V, 52 V and 76 V for 27 level MLI and 12 V, 24 V, 48 V and 82 V for 29 levels MLI output. A binary sequence DC voltage source is employed for the generation of 31 levels by the MLDCLI. The necessary DC sources values are 11 V, 22 V, 44 V and 88 V. Tables 5.1 and 5.2 show the modes of operations for all the configurations. Table 5.3 shows the specification of simulation and experimental setup of MLDCLI.

TABLE 5.3: Simulation and Experimental Specifications of MLDCLI Topology

Sl.	Parameters	Specifications
1	Voltage/Current/Power from all DC sources	41 V/1.3 A/150 W
2	Output RMS Voltage	110 V
3	Output RMS Current	3.6 A
4	Output Power	400 W
5	Output AC frequency	50 Hz
6	Simulation Software	MATLAB/Simulink 2015b
7	DC sources	Aplab (64 V/5 A, 4 nos)
8	IGBT	H15R1203 (1200 V/30 A, 12 nos)
9	Opto-isolator	TLP250 (12 nos)
10	Controller	AVR Atmega32 Microcontroller
11	Loading Rheostat	29 ohms/5 A
12	Inductive Load	50 mH/5 A
13	Power Quality Analyzer	Fluke 435b

5.3 Simulation Results

In this section, the Equal-Phase Method (EPM), Half-Equal Phase Method (HEPM), Selective Harmonics Elimination (SHE), Nearest Level Modulation (NLM) and modified Nearest Level Modulation (mNLM) schemes are simulated on MLDCLI topology, and the results are compared.

5.3.1 Single-phase Symmetric Configuration

Fig. 5.2 shows the 9-level single-phase simulated output voltage and current waveforms for all the five modulation schemes for the symmetric configuration of MLDCLI topology using R load. The mNLM, NLM and SHE schemes have similar looking waveforms due to smaller difference in the switching angles, but the HEPM and EPM schemes have a more significant difference in the switching angles, and hence their waveform appears smaller. Fig. 5.3 shows the Fast Fourier Transform (FFT) analysis (fundamental voltage and voltage THD) of the output voltage for all the five modulation schemes.

The simulation results are tabulated in Table 5.4. It can be seen that, in all the inverter parameters, the mNLM scheme has performed better than the other modulation schemes. The voltage THD (7.67 %) and current THD (7.67 %) of mNLM are least among all the modulation schemes. The RMS voltage (115.3 V) and RMS current (3.50 A) is also the highest among all the other modulation schemes. Since the RMS voltage and RMS current using mNLM is highest, the output power (401 W) is also the highest among all the other modulation schemes.

TABLE 5.4: Performance Comparison of Modulation Schemes with R Load on Single-phase MLDCLI Topology—Simulation

Inverter Parameters	EPM	HEPM	SHE	NLM	mNLM
Voltage THD (%)	24.81	21.41	8.23	8.35	**7.67**
Current THD (%)	24.81	21.41	8.23	8.35	**7.67**
RMS Voltage (V)	83.3	93.0	112.7	111.4	**115.3**
RMS Current (A)	2.52	2.82	3.41	3.38	**3.50**
Distortion Factor	0.9420	0.9561	0.9932	0.9930	**0.9941**
Output Power (W)	198	251	382	374	**401**

Fig. 5.4 shows the 9-level single-phase simulated output voltage and current waveforms for all the five modulation schemes for the symmetric configuration of MLDCLI topology using RL load. Figs. 5.5 and 5.6 show the FFT analysis of the output voltage and output current for all the five modulation

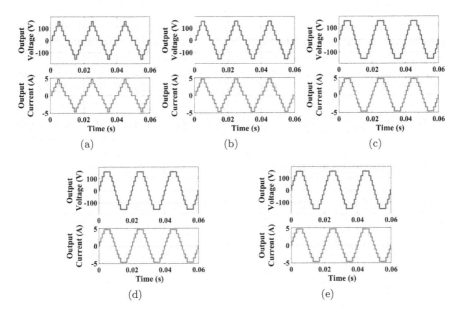

FIGURE 5.2: Simulated Output Voltage and Current Waveforms of Single-phase 9-Level MLDCLI with R Load using (a) EPM, (b) HEPM, (c) SHE, (d) NLM and (e) mNLM

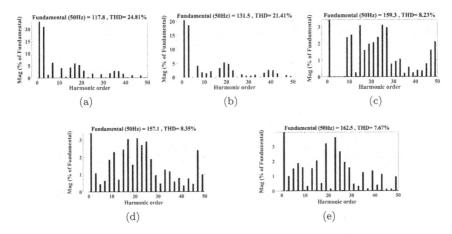

FIGURE 5.3: Simulated Output Voltage THD of Single-phase 9-Level MLD-CLI with R Load using (a) EPM, (b) HEPM, (c) SHE, (d) NLM and (e) mNLM

schemes. The FFT analysis shows the fundamental voltage, voltage THD and current THD using each modulation scheme.

The simulation results are tabulated in Table 5.5 where voltage THD, current THD, RMS voltage, RMS current, power factor and output power are the inverter parameters considered for comparison. It can be seen that, in all the inverter parameters, the mNLM scheme has performed better than the other modulation schemes. The voltage THD (7.89 %) and current THD (1.28 %) of mNLM are the least among all the modulation schemes. The RMS voltage (115.3 V) and RMS current (3.50 A) is also the highest among all modulation schemes. Since the RMS voltage and RMS current using mNLM is the highest, the output power (355 W) is also the highest. The next subsection deals with the various asymmetrical configurations of MLDCLI topology and the associated simulation results and performance analysis comparison of mNLM with other modulation schemes.

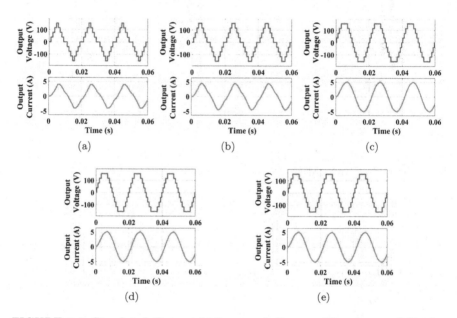

FIGURE 5.4: Simulated Output Voltage and Current Waveforms of Single-phase 9-Level MLDCLI with RL Load using (a) EPM, (b) HEPM, (c) SHE, (d) NLM and (e) mNLM

5.3.2   Single-phase Asymmetric Configurations

Fig. 5.7 presents the simulation output using mNLM for the 9-, 17-, 23- and 31-level configuration. It can be observed that, with the increase in the number of levels, the MLI output is approaching a sinewave shape.

Fig. 5.8 shows the FFT analysis of the four configurations. The fundamental voltage was found to be 162.5 V, 159 V, 158 V and 156.9 V for 9-, 17-, 23- and 31-level configuration respectively. The voltage THD was found to be 7.67%, 3.47%, 1.87% and 1.15% for the four configurations.

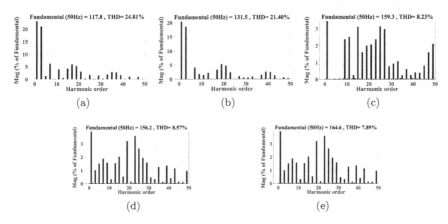

FIGURE 5.5: Simulated Output Voltage THD of Single-phase 9-Level MLD-CLI with RL Load using (a) EPM, (b) HEPM, (c) SHE, (d) NLM and (e) mNLM

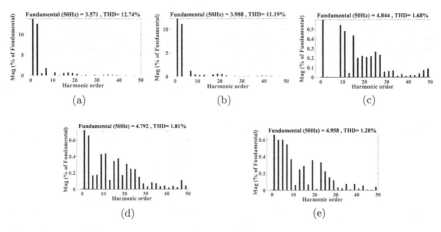

FIGURE 5.6: Simulated Output Current THD of Single-phase 9-Level MLD-CLI with RL Load using (a) EPM, (b) HEPM, (c) SHE, (d) NLM and (e) mNLM

Fig. 5.9 shows a comparison of simulation results between mNLM and NLM for four different inverter parameters, namely voltage THD, RMS voltage, RMS current and output power. The following inferences can be concluded:

1. An appreciable difference in the voltage THD between mNLM and NLM is seen. There is a minimum difference of 0.25 % for all the configurations.

2. There is a noticeable increase in the RMS voltage by an average of $3\ V_{rms}$ on mNLM compared to NLM scheme.

TABLE 5.5: Performance Comparison of Modulation Schemes with RL load on Single-phase MLDCLI Topology—Simulation

Inverter Parameters	EPM	HEPM	SHE	NLM	mNLM
Voltage THD (%)	24.81	21.40	8.23	8.57	**7.89**
Current THD (%)	12.74	11.19	1.68	1.81	**1.28**
RMS Voltage (V)	83.3	93.0	112.7	111.5	**115.3**
RMS Current (A)	2.52	2.82	3.43	3.39	**3.50**
Power Factor	0.8488	0.8568	0.8786	0.8783	**0.8788**
Output Power (W)	178	225	340	332	**355**

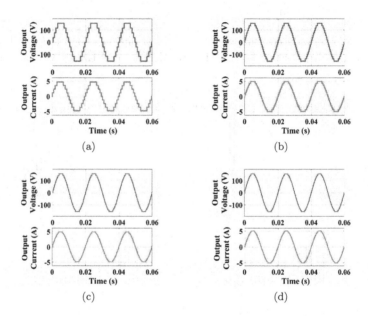

FIGURE 5.7: Simulated Output Voltage and Current Waveforms of Single-phase MLDCLI with R Load using mNLM for (a) 9-Level, (b) 17-Level, (c) 23-Level and (d) 31-Level

3. The RMS current has increased with the use of the proposed scheme.

4. The increase in the RMS voltage and RMS current together raised the output power to a higher level. There is visible increase in the output power.

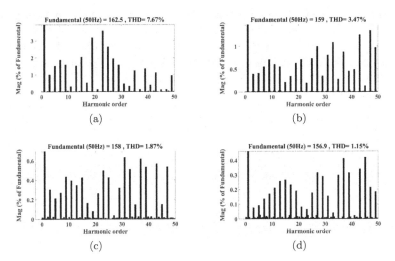

FIGURE 5.8: Simulated Output Voltage THD of Single-phase MLDCLI with R load using mNLM for (a) 9-Level, (b) 17-Level, (c) 23-Level and (d) 31-Level

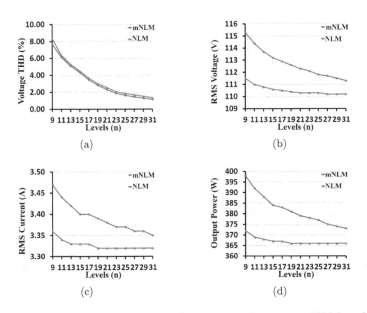

FIGURE 5.9: Simulated Parameters Comparison between mNLM and NLM (a) Voltage THD, (b) RMS Voltage, (c) RMS Current and (d) Output Power

5.3.3   Three-phase Symmetric Configuration

The simulation of single-phase symmetric configuration of MLDCLI topology is extended to three-phase star-connected configuration. The voltage and current output for the R and RL load are shown in Fig. 5.10. The simulation results are tabulated in Tables 5.6 and 5.7 for R and RL load respectively, where the proposed scheme is found to be effective among all the other modulation schemes.

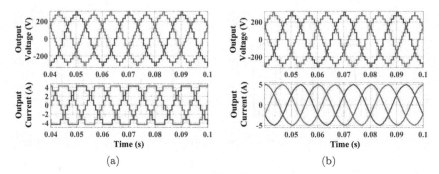

(a)   (b)

FIGURE 5.10: 3 ϕ V and I Output Waveforms using mNLM with (a) R Load and (b) RL Load

TABLE 5.6: Performance Comparison of Modulation Schemes with R Load on Three-phase MLDCLI Topology—Simulation

Inverter Parameters	EPM	HEPM	SHE	NLM	mNLM
Voltage THD (%)	11.54	9.92	6.30	6.33	**6.18**
Current THD (%)	24.90	21.48	8.36	8.50	**7.79**
RMS Voltage (V)	146.8	164.2	198.8	196.4	**203.4**
RMS Current (A)	2.31	2.58	3.12	3.09	**3.19**
Distortion Factor	0.9639	0.9729	0.9945	0.9944	**0.9950**
Output Power (W)	327	412	617	604	**646**

TABLE 5.7: Performance Comparison of Modulation Schemes with RL Load on Three-phase MLDCLI Topology—Simulation

Inverter Parameters	EPM	HEPM	SHE	NLM	mNLM
Voltage THD (%)	11.46	9.98	6.27	6.19	**6.10**
Current THD (%)	12.75	11.20	1.18	1.28	**1.00**
RMS Voltage (V)	146.9	163.8	198.7	196.4	**203.3**
RMS Current (A)	2.52	2.81	3.41	3.37	**3.49**
Power Factor	0.8689	0.8718	0.8799	0.8799	**0.8800**
Output Power (W)	322	401	596	582	**624**

5.4 Experimental Results

The setup consists of four DC sources, an MLDCLI main circuit, an AVR programmer, an Atmega32 microcontroller, a load and a Fluke 435 power quality analyzer. Four variable DC sources are employed to operate all the eleven configurations. The MLDCLI circuit was fabricated on a printed circuit board with a separate board for H-bridge. Port B of Atmega32 microcontroller was used for the generation of gate pulses. TLP250 acts as the driver circuit to fire the gate pulses of the semiconductor switches. Fluke 435b power quality analyzer was used to measure the voltage and current THD of the AC waveforms.

This section discusses the experimental results of the Equal-Phase Method (EPM), Half-Equal Phase Method (HEPM), Selective Harmonics Elimination (SHE), Nearest Level Modulation (NLM) and modified Nearest Level Modulation (mNLM) schemes on MLDCLI topology.

5.4.1 Single-phase Symmetric Configuration

Fig. 5.11 shows the 9-level single-phase experimental output voltage and current waveforms for all the five modulation schemes for the symmetric configuration of MLDCLI topology using R load. The EPM and HEPM schemes offer a poor quality AC waveform, and hence, the associated RMS voltage and RMS current were found to be reduced. On the other hand, the SHE, NLM, and mNLM schemes offered an excellent quality waveform with acceptable voltage THD and current THD values. But the mNLM scheme was observed to provide the highest RMS voltage and RMS current with least voltage THD.

Fig. 5.12 shows the FFT analysis of the output voltage for all the five modulation schemes. The analysis shows the fundamental voltage and voltage THD obtained using each modulation scheme.

The experimental results are tabulated in Table 5.8 where the inverter parameters for comparison considered are voltage THD, current THD, RMS voltage, RMS current, distortion factor and output power. It can be seen that, in all the inverter parameters, the mNLM scheme has performed better than the other modulation schemes. The voltage THD (7.7 %) and current THD (7.7 %) of mNLM are the least among all the modulation schemes. The RMS voltage (109.68 V) and RMS current (3.6 A) is also the highest. Since the RMS voltage and RMS current using mNLM is the highest, the output power (393 W) is also the highest among all the modulation schemes.

Fig. 5.13 shows the 9-level single-phase experimental output voltage and current waveforms for all the five modulation schemes of the symmetric configuration of MLDCLI topology using RL load. Fig. 5.14 and Fig. 5.15 show the FFT analysis of the output voltage and output current for all the five

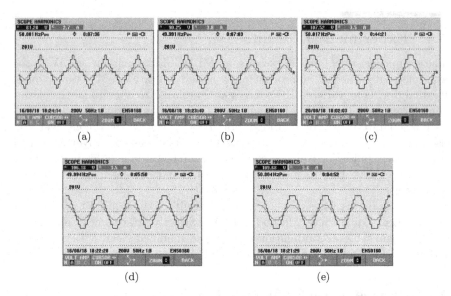

FIGURE 5.11: Hardware Output Voltage and Current Waveforms of Single-phase 9-Level MLDCLI with R Load using (a) EPM, (b) HEPM, (c) SHE, (d) NLM and (e) mNLM

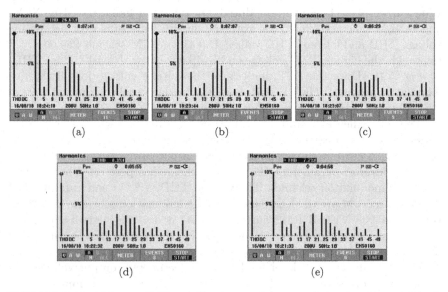

FIGURE 5.12: Voltage THD of Hardware Output of Single-phase 9-Level MLDCLI with R Load using (a) EPM, (b) HEPM, (c) SHE, (d) NLM, and (e) mNLM

TABLE 5.8: Performance Comparison of Modulation Schemes using R Load on Single-phase MLDCLI Topology—Hardware

Inverter Parameters	EPM	HEPM	SHE	NLM	mNLM
Voltage THD (%)	26.6	22.8	8.0	8.3	**7.7**
Current THD (%)	26.7	23.0	8.1	8.4	**7.7**
RMS Voltage (V)	81.28	90.25	107.52	106.18	**109.68**
RMS Current (A)	2.7	3.0	3.5	3.5	**3.6**
Distortion Factor	0.9336	0.9501	0.9935	0.9930	**0.9941**
Output Power (W)	205	257	374	369	**393**

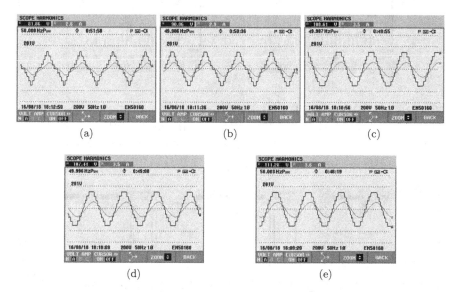

(a) (b) (c)

(d) (e)

FIGURE 5.13: Hardware Output Voltage and Current Waveforms of Single-phase 9-Level MLDCLI with RL Load using (a) EPM, (b) HEPM, (c) SHE, (d) NLM and (e) mNLM

modulation schemes. The FFT analysis shows the voltage THD and current THD using each modulation scheme.

Table 5.9 shows the experimental results where the inverter parameters for comparison considered are voltage THD, current THD, RMS voltage, RMS current, power factor and power. It can be seen that in all the inverter parameters, the mNLM scheme has performed better than the modulation schemes. The voltage THD (7.9 %) and current THD (1.4 %) of mNLM are the least. The RMS voltage (111.28 V), RMS current (3.6 A) and output power (352 W) are also the highest among all the schemes.

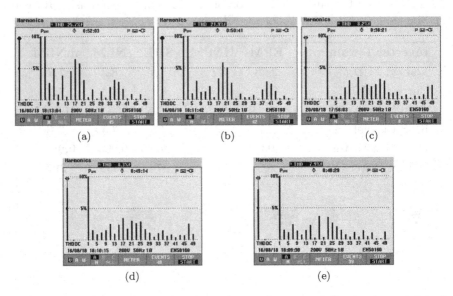

FIGURE 5.14: Voltage THD of Hardware Output of Single-phase 9-Level MLDCLI with RL Load using (a) EPM, (b) HEPM, (c) SHE, (d) NLM and (e) mNLM

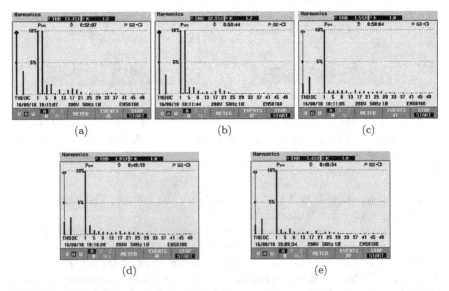

FIGURE 5.15: Current THD of Hardware Output of Single-phase 9-Level MLDCLI with RL Load using (a) EPM, (b) HEPM, (c) SHE, (d) NLM and (e) mNLM

TABLE 5.9: Performance Comparison of Modulation Schemes with RL Load on Single-phase MLDCLI Topology—Hardware

Inverter Parameters	EPM	HEPM	SHE	NLM	mNLM
Voltage THD (%)	25.7	21.9	8.2	8.3	**7.9**
Current THD (%)	13.4	12.1	1.5	1.9	**1.4**
RMS Voltage (V)	81.86	90.86	108.61	107.44	**111.28**
RMS Current (A)	2.6	2.9	3.5	3.5	**3.6**
Power Factor	0.8463	0.8550	0.8786	0.8785	**0.8788**
Output Power (W)	180	225	334	330	**352**

5.4.2 Single-phase Asymmetric Configurations

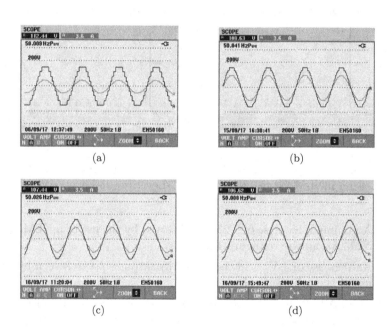

(a)

(b)

(c)

(d)

FIGURE 5.16: Hardware Output Voltage and Current Waveforms of Single-phase MLDCLI with R Load using mNLM for (a) 9-Level, (b) 17-Level, (c) 23-Level and (d) 31-Level

Fig. 5.16 presents the hardware voltage and current waveforms output using mNLM for the 9-, 17-, 23- and 31-level configuration. The RMS voltage was found to be 112.44 V, 108.63 V, 107.44 V and 106.62 V for the 9-, 17-, 23- and 31-level configuration respectively.

Fig. 5.17 shows the FFT analysis of the four configurations. The voltage THD was found to be 7.5%, 3.1%, 2.3% and 2.2% for the four configurations.

Fig. 5.18 shows the comparison of experimental results between mNLM and NLM for four different inverter parameters, namely, voltage THD, RMS voltage, RMS current and output power. The following conclusions are drawn:

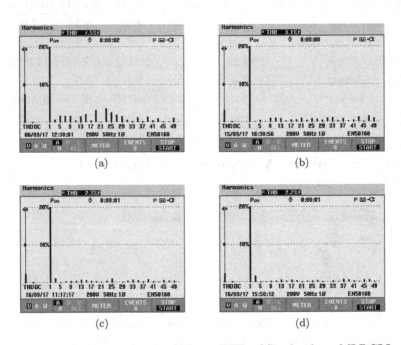

(a)   (b)

(c)   (d)

FIGURE 5.17: Hardware Output Voltage THD of Single-phase MLDCLI with R Load using mNLM for (a) 9-Level (b) 17-Level (c) 23-Level (d) 31-Level

1. A minimum increase of 0.25% is observed by the use of mNLM over NLM for all the configurations.

2. An increase of 4 V_{rms} in the AC RMS output voltage is seen by the use of the proposed scheme.

3. The RMS current has also shown an increase in the magnitude of mNLM compared to NLM.

4. A good increase in the RMS voltage and RMS current has reflected in a significant rise in the output power by the use of the mNLM over NLM.

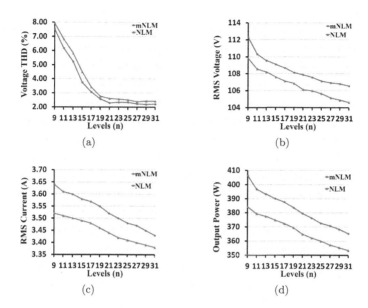

FIGURE 5.18: Hardware Parameters Comparison between mNLM and NLM (a) Voltage THD, (b) RMS Voltage, (c) RMS Current and (d) Output Power

5.5 Conclusion

An MLDCLI topology is presented which was able to generate MLI levels as high as 31-level from 9-level. The symmetrical configuration was investigated with all the LSF modulation schemes. A comparison between NLM and mNLM scheme for asymmetrical configuration was also presented. The CHBI topology was designed for a 400 W AC output system. The performance was studied using a MATLAB/Simulink 2015b environment and a hardware prototype of a single-phase system of the same demonstrated for validation of the effectiveness of the modulation scheme proposed. The simulation was performed for symmetric and asymmetric configurations and extended to a three-phase system of symmetric configuration. The simulation and hardware experiments were carried out for R and RL load.

6

Practical Implementation of MLIs in Power Conversion

This chapter provides a brief insight into the practical implementation of MLIs in the field of power conversion. MLIs have found application to promising fields in power electronics such as integration with Renewable Energy Sources (RES) and Electric Vehicles (EV).

6.1 MLI applications to Renewable Energy Sources (RES)

Renewable Energy Sources (RES) are seen as a way to meet the ever-increasing demand for power and energy to the whole of humanity. The motivation behind the use of RES is that it offers an alternative to the use of fossil fuels which are the main reason for all the forms of pollutions throughout the world. The increasing price of the fuels also adds to the economic problems. The RES provides a clean source of energy without causing any pollution. The second most important reason for employing RES is their decentralized nature of production. A conventional power system needs to carry power from the generation system to transmission system followed by the distribution system. It requires massive investment in the form of cable wires, transformers and other equipment together with manpower. The maintenance is another aspect of the power system. Due to the decentralized nature of RES, all the problems associated with a conventional power system is avoided.

There are various forms of RES, out of which solar energy and wind energy are the most prominent sources of renewable energy. By employing Photovoltaic (PV) Panels, solar energy is converted to electrical energy. On the other hand, the electrical turbine system is used to generate electricity from wind energy. For the solar energy and wind energy system, power electronics is a critical aspect to convert the power from one form to the another. The power generated from the PV panel and wind mill are often varying in nature, and hence, it is vital to maintain the voltage levels according to the varying nature of input power.

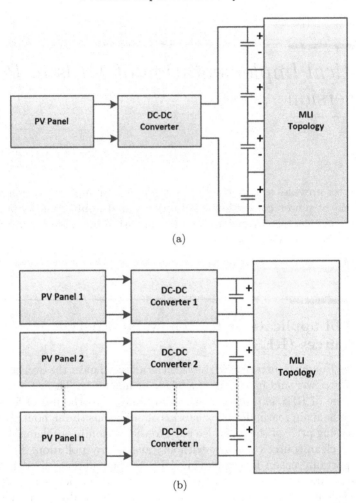

(a)

(b)

FIGURE 6.1: PV panels connection to MLIs (a) Non-isolated and (b) Isolated

The Fig. 6.1 shows the two ways in which the PV panels could be connected to the MLI topologies. The first configuration is a non-isolated type where a single PV source is split into multiple sources and fed as input to a MLI topology. This type of configuration is usually found in DC-MLI and FC-MLI topologies. As all the DC voltage sources to the MLI are equal, it is an example of symmetric type MLI topology. The second configuration is the isolated type where multiple PV panels are used as inputs to the MLI topology. The isolated type can operate the MLI in asymmetric mode, thereby generating an increased number of levels. For practical applications such as PV, where the PV panels face shading conditions, it is preferred to employ isolated type. This allows for individual voltage control of each source. This leads to a redundant operation of MLIs. Thus, the isolated type is gaining increased preference over non-isolated because of numerous advantages.

In both the types a PV panel is connected to a DC-DC power electronics converter and then fed as the input source to MLIs. The DC-DC converter is a critical component whose primary role is to maintain the DC voltage to a fixed magnitude. The voltage from PV panel is subjected to vary due to the variations in the insolation level from the sun. This is because of the shading effect caused due to moving clouds, shadows of nearby objects, bird droppings, leaves falling and so on. By varying the duty cycle, the DC-DC converter maintains the output voltage to a constant level. The DC-DC converter also performs the maximum power point tracking (MPPT) algorithm to attain maximum power.

The MLIs are used for high power applications in a power system where power is transmitted in kiloWatt. A commercially available single PV panel can generate power in the range of 100 W to 500 W. The PV panels are required to be connected in series/parallel combinations to get sufficient voltage and current. Hence, there are various available combinations of PV panels. For gaining an increased voltage, the PV panels are required to be connected in series. On the other hand, to increase the current rating, the panels are required to be connected in parallel.

The following are the four types of PV panel configurations depending on the connections of each panel and the placement of power electronics converters [50].

1. *Central Inverter System*: A Central Inverter System (CIS) consists of several series and parallel combinations of PV panels. By combining the required number of PV panels in series to achieve the inverter voltage, the need for a voltage boosting converter is avoided. However, a boost DC-DC converter is still employed to maintain the inverter voltage when there is a drop in the PV voltage due to partial shading conditions. This is a conventional type of system often used in many residential buildings. The major drawback with CIS is that a single PV panel shaded will bring down the net output power of CIS to a drastically low net output power.

2. *String Inverter System*: A String Inverter System (SIS) consists of the required number of PV panels added in series and then connected to an individual inverter. This leads to the need for relatively lower power ratings of the inverter. The string diode is also avoided here. Each string also has a DC-DC converter which carried out MPPT for each of the strings individually. Hence, a higher power is obtained, and the efficiency of the whole system is improved.

3. *Multi-String Inverter System*: A Multi-string Inverter System (MIS) has several single string PV panels connected with individual

DC-DC converters for MPPT. However, there is only a single inverter to which all the outputs of the DC-DC converters are connected. This leads to reducing the unnecessary need for multiple inverters as in SIS. For a low power application, the MIS is suggested for use.

4. *Module Integrated Converter*: Module Integrated Converter (MIC) is an upcoming field of converter topologies used with an individual PV panel. The MIC requires a high gain of conversion from PV output to direct inverter input. A high level of efficiency and better output power are the results of MIC. The MPPT and inverter system are embedded within a single power electronic converter to form the MIC. But it requires a higher initial investment to set up.

The Fig. 6.2 shows two ways in which the wind turbine system can be integrated with MLIs and much similar to the PV system. The power from the wind turbine is generated in the form of AC power. This AC power is first converted to DC power with the use of a three-phase diode bridge rectifier. An uncontrolled three-phase rectifier is a most straightforward and useful method to convert AC power into DC power. There is no complex control circuitry involved, and hence, the power conversion is straightforward. The DC power is then fed to the DC-DC converter which is then transferred to MLI topology. The DC-DC converter can be in the form of a boost converter which will provide the required step-up voltage to meet the voltage rating of the grid.

A single wind turbine system may be used as a single source of DC power. As such a conventional DC-MLI with a higher number of levels can be implemented. It also requires proper capacitor balancing to get a smooth MLI output. The other way is the use of a multi winding transformer to provide isolated inputs to the MLI topology. The isolated inputs can be used for MLI topologies such as CHB and MLDCLI. The wind energy is also considered to be varying in nature, but unlike solar energy, wind energy is available even during the night hours. Hence, wind energy can be expected throughout the day.

A comparison of the recent developments in the application of MLIs to RES is shown in Table 6.1. The table provides a summary of the type of renewable energy source used with the MLI topology employed. It also indicates the type of modulation scheme used.

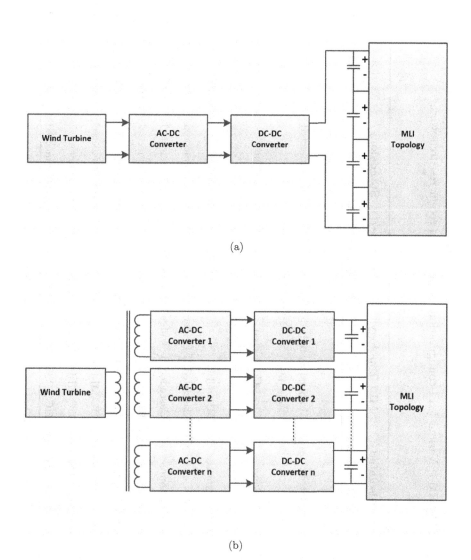

(a)

(b)

FIGURE 6.2: Wind turbine connection to MLIs (a) Single DC source and (b) Multiple DC source

TABLE 6.1: Various MLI configurations with RES in recent research works

Sl.	Ref.	Type of RES	Non-isolated/ Isolated Sources	MLI Topology	Single Phase/ Three Phase	Standalone/ Grid-Connected	Modulation Scheme	MLI Levels
1	[51]	Solar	Non-isolated Sources	Modified Switched Capacitor	Single-Phase	Grid-Connected	HSF	Five
2	[46]	Solar	Isolated Sources	Cross-Connected Sources	Single-Phase	Standalone	HSF	Five/ Seven
3	[52]	Solar	Isolated Sources	MLDCLI	Single-Phase	Standalone	LSF	Nine
4	[53]	Solar	Isolated Sources	Cascaded H-Bridge	Three-Phase	Grid-Connected	HSF	Seven
5	[54]	Solar	Isolated Sources	Cascaded H-Bridge	Single-Phase	Grid-Connected	HSF	Nine
6	[55]	Wind	Non-isolated Sources	DC-MLI	Three-Phase	Grid-Connected	HSF	Five
7	[56]	Wind	Isolated Sources	Cascaded H-Bridge	Single-Phase	Standalone	HSF	Eleven
8	[57]	Wind	Non-isolated Sources	DC-MLI	Three-Phase	Grid-Connected	HSF	Five
9	[58]	Wind	Non-isolated Sources	DC-MLI	Single-Phase	Standalone	LSF	Nine
10	[59]	Wind	Non-isolated Sources	DC-MLI	Three-Phase	Grid-Connected	HSF	Three

6.2 MLI applications in Electric Vehicles (EVs)

The development of Electric Vehicles (EVs) are highly dependent on the future of power electronics technology. An EV compared to an ICE engine has almost nil mechanical part. A lot of power conversion takes place in the EV ecosystem which includes the powertrain, car lightning and power to other car accessories. Therefore, there are varying types of load which requires different voltage, current and power ratings. It requires different power converter topologies for each load.

For driving the powertrain of the EV, induction motors are the primary source of traction. To drive the induction motor, a three phase inverter technology is employed. With the advent of MLIs, various MLI topologies are used in the three phase inverter system as they offer a higher VA rating. The efficient operation of a induction motor also requires a higher voltage rating, lower THD and lower EMI features. The MLI also offers a transformer less operation which greatly reduces the size and weight of the EVs. A reduction in weight allows for better fuel consumption and lasting fuel tank. The figure 6.3 shows the use of a three phase CHBI in EV traction.

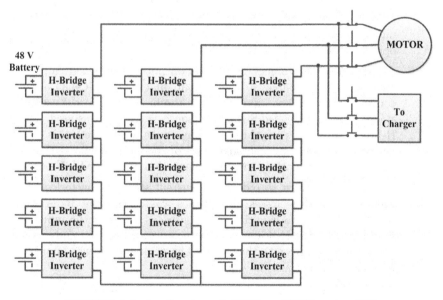

FIGURE 6.3: Application of CHBI to EV Traction [60]

The use of MLIs in the Hybrid Electric Vehicle (HEV) is also becoming popular. In a parallel HEV configuration, in addition to the combustion engine, the combination of batteries and ultra capacitors would provide a "power assist" through an inverter to the summing gear which drives the EV motor [61].

In the recent years, the applications of MLI to EVs have taken a new dimension in the form of different topologies and control schemes. There are also reports of newer generation of semiconductor devices being employed in the MLIs. Some of the recent reports on MLI development are as follows:

1. A T-type MLI structure is used by Ahmed Sheir et al. in achieving a bidirectional power flow to charge the batteries. A DC link capacitor is also used in the DC-DC converter as a way to achieve capacitor voltage balancing. The converter structure has utilized a capacitor and two additional power switches which helps in avoiding the use of any additional sensor or control loops. A 1 kW prototype of the converter was developed and tested with successfully [62].

2. Fengqi Change et. al. had taken up three different MLI structure and compared their performance. An IGBT based inverter, a SiC based inverter and a CHB with Si based switches are used in the work. The Si based CHB was found to have offered a better efficiency and a lesser cost among the compared MLI topologies [63].

3. A fault detection algorithm was developed by Anton Kersten et. al. for a three level DC-MLI with a connected neutral point. An adaptive SVM algorithm was used which can help in driving the powertrain with available maximum power to a near by station under a fault condition. This operation is referred to as "limp mode" operation of the electric vehicle [64].

4. An MLI topology to generate 7 level output was designed employing a single DC source. A single input dual output DC-DC boost converter was used as input which is the first stage. In the second stage, a diode by-passed Switch DC source converter was used. The H-bridge inverter is the third and final stage where the DC is converted to AC output. The advantage of such a converter is that a lesser number of battery in series connection are needed [65].

5. An interesting work to reduce the current stress on batteries and thereby reduce the losses in batteries was carried out by Anton Kersten et. al. Two capacitors were used as filters at the input of H-bridge cell. The H-bridge cells were cascaded together and a star connection was given to the motor. The peak current was found to have reduced by 5% to 20% [66].

The above reports give a brief overview of the research trends in the MLI applications to EVs. However, there exists a lot of recent works on EVs employing MLIs.

6.3 Future Scope of LSF Modulation Schemes for RES and EV applications

Given the advantages LSF modulation schemes carry over HSF modulation schemes, there is a good amount of scope to realize the benefit of LSF modulation on MLIs with RES and EVs. Some of the aspects over which the LSF schemes can be researched upon are the following:

1. *Optimizing switching angles for variations in PV and wind output*: The PV and wind output are always varying due to change in weather conditions affecting voltage and current output. For MLIs, it is necessary to maintain a constant output voltage. Even with the best use of DC-DC converters, the duty cycle cannot be operated at extreme. Hence, even for a decreased voltage magnitude, the switching angles can be suitably adjusted to maintain the THD to minimum. In this case, the output RMS voltage is reduced but it does not effect the quality of AC output.

2. *Redundant MLI topologies*: When integrating MLIs with various renewable energy sources or batteries in EVs where multiple input sources are employed, there are possibilities of blackout of any one single source. During blackout of a single source, the number of output MLI levels fall from the original no. of levels. Hence, it is very important for the MLI controller to change the switching angle to adjust to the reduced no. of MLI levels.

3. *LSF modulation schemes for dynamic loads*: Loads in a real world are always dynamic. The power system and EVs undergoes a constant load change and hence, it is important for the MLI to meet the load change. This is especially true when a battery system is employed with bidirectional power flow at the input of MLI. The battery pack needs an intelligent algorithm which should be well integrated with MLI modulation scheme to cater to the needs of the power system.

4. *Hybrid LSF schemes*: The conventional LSF modulation schemes can also be combined with each other to realize the benefit of each scheme. One such good combination is NLM with SHE, which offers a simple NLM scheme with the harmonics elimination advantage of SHE scheme. The advanced variation of a conventional scheme can also be used with other schemes to realize an added benefit.

The above are some of the important scopes to which an LSF scheme can undergo a sufficient amount of research in near future. These points offer ample opportunities to carry forward the research in the area of LSF modulation schemes.

Bibliography

[1] José Rodríguez, Jih Sheng Lai, and Fang Zheng Peng. Multilevel inverters: A survey of topologies, controls, and applications. *IEEE Transactions on Industrial Electronics*, 49(4):724–738, 2002.

[2] Amarendra Edpuganti and Akshay Kumar Rathore. A survey of low switching frequency modulation techniques for medium-voltage multilevel converters. *IEEE Transactions on Industry Applications*, 51(5):4212–4228, 2015.

[3] Rakesh Kumar, Deepa Thangavelusamy, Sanjeevikumar Padmanaban, and D P Kothari. A guide to Nearest Level Modulation and Selective Harmonics Elimination modulation scheme for multilevel inverters. In *2019 Innovations in Power and Advanced Computing Technologies (i-PACT)*, number March, pages 1–8, 2019.

[4] A Rakesh Kumar and T. Deepa. Multilevel Inverters: A Review of Recent Topologies and New Modulation Techniques. In *2018 International Conference on Recent Trends in Electrical, Control and Communication (RTECC)*, pages 196–203. IEEE, mar 2018.

[5] Manigandan Thathan and Albert Alexander. Modelling and analysis of modular multilevel converter for solar photovoltaic applications to improve power quality. *IET Renewable Power Generation*, 9(1):78–88, jan 2015.

[6] Minjie Chen, Khurram K. Afridi, and David J. Perreault. A multilevel energy buffer and voltage modulator for grid-interfaced microinverters. *IEEE Transactions on Power Electronics*, 30(3):1203–1219, 2015.

[7] Hamed Nademi, Anandarup Das, Rolando Burgos, and Lars E Norum. A New Circuit Performance of Modular Multilevel Inverter Suitable for Photovoltaic Conversion Plants. *IEEE Journal of Emerging and Selected Topics in Power Electronics*, 4(2):393–404, 2016.

[8] Viju Nair R, K Gopakumar, and Leopoldo G Franquelo. A Very High Resolution Stacked Multilevel Inverter Topology for Adjustable Speed Drives. *IEEE Transactions on Industrial Electronics*, 65(3):2049–2056, mar 2018.

[9] A S Aneesh Kumar, Gautam Poddar, and P. Ganesan. Control strategy to naturally balance hybrid converter for variable-speed medium-voltage drive applications. *IEEE Transactions on Industrial Electronics*, 62(2):866–876, 2015.

[10] Mohamed Z. Youssef, Konrad Woronowicz, Kunwar Aditya, Najath Abdul Azeez, and Sheldon S. Williamson. Design and development of an efficient multilevel DC/AC traction inverter for railway transportation electrification. *IEEE Transactions on Power Electronics*, 31(4):3036–3042, 2016.

[11] Rajasegharan V.V., Premalatha L., and Rengaraj R. Modelling and controlling of PV connected quasi Z-source cascaded multilevel inverter system: An HACSNN based control approach. *Electric Power Systems Research*, 162(April):10–22, sep 2018.

[12] Ankit, Sarat Kumar Sahoo, Sukruedee Sukchai, and Franco Fernando Yanine. Review and comparative study of single-stage inverters for a PV system. *Renewable and Sustainable Energy Reviews*, 91(March 2017):962–986, aug 2018.

[13] Mehran Sabahi, Mohammad Farhadi Kangarlu, and Ebrahim Babaei. Dynamic voltage restorer based on multilevel inverter with adjustable dc-link voltage. *IET Power Electronics*, 7(3):576–590, mar 2014.

[14] Natarajan Prabaharan and Kaliannan Palanisamy. A comprehensive review on reduced switch multilevel inverter topologies, modulation techniques and applications. *Renewable and Sustainable Energy Reviews*, 76(April):1248–1282, 2017.

[15] Muhammad H Rashid. *Power Electronics Handbook*. Elsevier Ltd, 3rd edition, 2011.

[16] T. A. Meynard and H. Foch. Multi-level conversion: high voltage choppers and voltage-source inverters. In *Power Electronics Specialists Conference, 1992. PESC '92 Record., 23rd Annual IEEE*, pages 397–403, 1992.

[17] José Rodríguez, Luis Morán, Pablo Correa, and Cesar Silva. A Vector Control Technique for Medium-Voltage Multilevel Inverters. *IEEE Transactions on Industrial Electronics*, 49(4):882–888, 2002.

[18] Marcelo A. Perez, Steffen Bernet, Jose Rodriguez, Samir Kouro, and Ricardo Lizana. Circuit Topologies, Modeling, Control Schemes, and Applications of Modular Multilevel Converters. *IEEE Transactions on Power Electronics*, 30(1):4–17, jan 2015.

[19] Georgios Konstantinou, Josep Pou, Gabriel J Capella, Kejian Song, Salvador Ceballos, and Vassilios G Agelidis. Interleaved Operation of Three-Level Neutral Point Clamped Converter Legs and Reduction of Circulating Currents Under SHE-PWM. *IEEE Transactions on Industrial Applications*, 63(6):3323–3332, 2016.

[20] Sheng-Feng Wu and Chung-Ming Young. Selective harmonic elimination in multi-level inverter with zig-zag connection transformers. *IET Power Electronics*, 7(4):876–885, 2014.

[21] Mohammad Sharifzadeh, Hani Vahedi, Abdolreza Sheikholeslami, Philippe-Alexandre Labbe, and Kamal Al-Haddad. Hybrid SHM-SHE Modulation Technique for Hybrid SHM-SHE Modulation Technique for Four-Leg NPC Inverter with DC Capacitors. *IEEE Transactions on Industrial Electronics*, 62(8):4890–4899, 2015.

[22] C Bharatiraja, Seenithangam Jeevananthan, Ramachandran Latha, and V Mohan. Vector selection approach-based hexagonal hysteresis space vector current controller for a three phase diode clamped MLI with capacitor voltage balancing. *IET Power Electronics*, 9(7):1350–1361, 2016.

[23] Arul Rahul Sanjeevan, R. Sudharshan Kaarthik, K. Gopakumar, P.P. Rajeevan, Jose I. Leon, and Leopoldo G. Franquelo. Reduced common-mode voltage operation of a new seven-level hybrid multilevel inverter topology with a single DC voltage source. *IET Power Electronics*, 9(3):519–528, 2016.

[24] Ui Min Choi, Frede Blaabjerg, and Kyo Beum Lee. Reliability improvement of a T-type three-level inverter with fault-tolerant control strategy. *IEEE Transactions on Power Electronics*, 30(5):2660–2673, 2015.

[25] Jin-Sung Choi and Feel-Soon Kang. Seven-level PWM inverter employing series-connected capacitors paralleled to a single DC voltage source. *IEEE Transactions on Industrial Electronics*, 62(6):3448–3459, 2015.

[26] Jiandong Duan, Fengjiang Wu, and Fan Feng. Modified single-carrier multilevel sinusoidal pulse width modulation for asymmetrical insulated gate bipolar transistor-clamped grid-connected inverter. *IET Power Electronics*, 8(8):1531–1541, 2015.

[27] Uthandipalayam Subramaniyam Ragupathy, Ramasami Uthirasamy, and Venkatachalam Kumar Chinnaiyan. Structure of boost DC-link cascaded multilevel inverter for uninterrupted power supply applications. *IET Power Electronics*, 8(11):2085–2096, 2015.

[28] Mehdi Hajizadeh and Seyed Hamid Fathi. Selective harmonic elimination strategy for cascaded H-bridge five-level inverter with arbitrary power sharing among the cells. *IET Power Electronics*, 9(1):95–101, 2016.

[29] Sze Sing Lee, Bing Chu, Nik Rumzi Nik Idris, Hui Hwang Goh, and Yeh En Heng. Switched-battery boost-multilevel inverter with GA optimized SHEPWM for standalone application. *IEEE Transactions on Industrial Electronics*, 63(4):2133–2142, 2016.

[30] Hui Zhao, Tian Jin, Shuo Wang, and Liang Sun. A real-time selective harmonic elimination based on a transient-free inner closed-loop control for cascaded multilevel inverters. *IEEE Transactions on Power Electronics*, 31(2):1000–1014, 2016.

[31] Mohamad Reza Banaei, Alireza Dehghanzadeh, and Aida Baghbany Oskouei. Extended switching algorithms based space vector control for five-level quasi-Z-source inverter with coupled inductors. *IET Power Electronics*, 7(6):1509–1518, 2014.

[32] V Karthikeyan, A Rakesh Kumar, and V Jamuna. A New Multilevel Inverter with BCD Topology and Reduction of Harmonics Using Sine Property. *Archives Des Sciences*, 66(4):712–719, 2013.

[33] T Deepa and A. Rakesh Kumar. A new asymmetric multilevel inverter with reduced number of switches and reduction of harmonics using Sine Property. In *2016 International Conference on Circuit, Power and Computing Technologies (ICCPCT)*, pages 1–8. IEEE, mar 2016.

[34] Shalini Tahunguriya, A Rakesh Kumar, and T Deepa. Multilevel Inverter with Reduced Number of Switches and Reduction of Harmonics. *Middle-East Journal of Scientific Research*, 24(S1):184–191, 2016.

[35] A Rakesh Kumar, Mahajan Sagar Bhaskar, Umashankar Subramaniam, Dhafer Almakhles, Sanjeevikumar Padmanaban, and Jens Bo-Holm Nielsen. An Improved Harmonics Mitigation Scheme for a Modular Multilevel Converter. *IEEE Access*, 7(Mmc):147244–147255, 2019.

[36] Emad Samadaei, Mohammad Kaviani, and Kent Bertilsson. A 13-levels Module (K-Type) with two DC sources for Multilevel Inverters. *IEEE Transactions on Industrial Electronics*, 66(7):5186–5196, 2019.

[37] Emad Samadaei, Abdolreza Sheikholeslami, Sayyed Asghar Gholamian, and Jafar Adabi. A Square T-Type (ST-Type) Module for Asymmetrical Multilevel Inverters. *IEEE Transactions on Power Electronics*, 33(2):987–996, feb 2018.

[38] Ravi Raushan, Bidyut Mahato, and K.C. Jana. Optimum structure of a generalized three-phase reduced switch multilevel inverter. *Electric Power Systems Research*, 157:10–19, apr 2018.

[39] Cheng-Lin Xiong, Xiao-Yun Feng, Fei Diao, and Xia-Jie Wu. Improved nearest level modulation for cascaded H-bridge converter. *Electronics Letters*, 52(8):648–650, apr 2016.

[40] Pengfei Hu and Daozhuo Jiang. A level-increased nearest level modulation method for modular multilevel converters. *IEEE Transactions on Power Electronics*, 30(4):1836–1842, 2015.

[41] Prafullachandra M. Meshram and Vijay B. Borghate. A Simplified Nearest Level Control (NLC) Voltage Balancing Method for Modular Multilevel Converter (MMC). *IEEE Transactions on Power Electronics*, 30(1):450–462, jan 2015.

[42] Ashraf Ahmed, Mohana Sundar Manoharan, Joung-hu Park, and Senior Member. An Efficient Single-Sourced Asymmetrical Cascaded Multilevel Inverter With Reduced Leakage Current Suitable for Single-Stage PV Systems. *IEEE Transactions on Energy Conversion*, 34(1):211–220, 2019.

[43] Raghavendra Reddy Karasani, Vijay Bhanuji Borghate, Prafullachandra M. Meshram, Hiralal Murlidhar Suryawanshi, and Sidharth Sabyasachi. A three-phase hybrid cascaded modular multilevel inverter for renewable energy environment. *IEEE Transactions on Power Electronics*, 32(2):1070–1087, 2017.

[44] Rekha Agrawal and Shailendra Jain. Multilevel inverter for interfacing renewable energy sources with low/medium- and high-voltage grids. *IET Renewable Power Generation*, 11(14):1822–1831, dec 2017.

[45] Ming Qi and Om. P. Malik. Apply STATCOM with a Novel Topology to the Power Sub Grid. In *2014 IEEE Electrical Power and Energy Conference*, pages 242–247. IEEE, nov 2014.

[46] Krishna Kumar Gupta and Shailendra Jain. Comprehensive review of a recently proposed multilevel inverter. *IET Power Electronics*, 7(3):467–479, 2014.

[47] Yi Hung Liao and Ching Ming Lai. Newly-constructed simplified single-phase multistring multilevel inverter topology for distributed energy resources. *IEEE Transactions on Power Electronics*, 26(9):2386–2392, 2011.

[48] A Rakesh Kumar and Deepa Thangavelusamy. A modified nearest level modulation scheme for symmetric and asymmetric configurations of cascaded H-bridge inverter. *International Journal of Electrical Engineering & Education*, pages 1–16, jun 2019.

[49] Ramkumar L Maurya and Mini Rajeev. Implementation of multilevel DC-link inverter for standalone application. In *2017 International Conference on Nascent Technologies in Engineering (ICNTE)*, pages 1–6, 2017.

[50] Jagabar Sathik Mohd.Ali and Vijayakumar Krishnaswamy. An assessment of recent multilevel inverter topologies with reduced power electronics components for renewable applications. *Renewable and Sustainable Energy Reviews*, 82(July):3379–3399, feb 2018.

[51] Naser Vosoughi, Seyed Hossein Hosseini, and Mehran Sabahi. A New Transformer-Less Five-Level Grid-Tied Inverter for Photovoltaic Applications. *IEEE Transactions on Energy Conversion*, 35(1):106–118, mar 2020.

[52] A. Rakesh Kumar, Deepa Thangavelusamy, Sanjeevikumar Padmanaban, and Dwarkadas P. Kothari. A Modified PWM Scheme to improve AC Power Quality for MLIs using PV Source. *International Journal of Power and Energy Systems*, 39(1):34–41, 2019.

[53] Rahul Sharma and Anandarup Das. Extended Reactive Power Exchange With Faulty Cells in Grid-Tied Cascaded H-Bridge Converter for Solar Photovoltaic Application. *IEEE Transactions on Power Electronics*, 35(6):5683–5691, jun 2020.

[54] Cheng Wang, Kai Zhang, Jian Xiong, Yaosuo Xue, and Wenxin Liu. An Efficient Modulation Strategy for Cascaded Photovoltaic Systems Suffering From Module Mismatch. *IEEE Journal of Emerging and Selected Topics in Power Electronics*, 6(2):941–954, jun 2018.

[55] Ahmed A. Hossam-Eldin, Emtethal Negm, Mohamed S. Elgamal, and Kareem M. AboRas. Operation of grid-connected DFIG using SPWM- and THIPWM-based diode-clamped multilevel inverters. *IET Generation, Transmission & Distribution*, 14(8):1412–1419, apr 2020.

[56] Sumit K. Chattopadhyay and Chandan Chakraborty. Full-Bridge Converter With Naturally Balanced Modular Cascaded H-Bridge Wave-shapers for Offshore HVDC Transmission. *IEEE Transactions on Sustainable Energy*, 11(1):271–281, jan 2020.

[57] Sherif M. Dabour, Ibrahim Masoud, and Essam M. Rashad. Carrier-Based PWM Technique for a New Six-to-Three-Phase Multilevel Matrix Converter for Wind-Energy Conversion Systems. In *2019 IEEE Conference on Power Electronics and Renewable Energy (CPERE)*, pages 412–417. IEEE, oct 2019.

[58] Akram Mohammed, Nashiren Farzilah Mailah, Mohd Amran Mohd Radzi, Mohd Khair Hassan, Yuan Kang Wu, and Yi-Liang Hu. Modulation Index Sets of Low Switching Frequency Multilevel Inverters for Wind Generation System. In *2018 IEEE PES Asia-Pacific Power and Energy Engineering Conference (APPEEC)*, volume 2018-Octob, pages 246–251. IEEE, oct 2018.

[59] Venkata Yaramasu and Bin Wu. Predictive Control of a Three-Level Boost Converter and an NPC Inverter for High-Power PMSG-Based Medium Voltage Wind Energy Conversion Systems. *IEEE Transactions on Power Electronics*, 29(10):5308–5322, oct 2014.

[60] Leon M. Tolbert, Fang Z. Peng, and Thomas G. Habetler. Multilevel inverters for electric vehicle applications. *IEEE Workshop on Power Electronics in Transportation*, pages 79–84, 1998.

[61] Leon M. Tolbert, Fang Zheng Peng, Tim Cunnyngham, and John N. Chiasson. Charge balance control schemes for cascade multilevel converter in hybrid electric vehicles. *IEEE Transactions on Industrial Electronics*, 49(5):1058–1064, 2002.

[62] Ahmed Sheir, Mohamed Z. Youssef, and Mohamed Orabi. A novel bidirectional T-type multilevel inverter for electric vehicle applications. *IEEE Transactions on Power Electronics*, 34(7):6648–6658, 2019.

[63] Fengqi Chang, Olga Ilina, Markus Lienkamp, and Leon Voss. Improving the Overall Efficiency of Automotive Inverters Using a Multilevel Converter Composed of Low Voltage Si mosfets. *IEEE Transactions on Power Electronics*, 34(4):3586–3602, 2019.

[64] Anton Kersten, Karl Oberdieck, Andreas Bubert, Markus Neubert, Emma Arfa Grunditz, Torbjorn Thiringer, and Rik W. De Doncker. Fault detection and localization for limp home functionality of three-level NPC inverters with connected neutral point for electric vehicles. *IEEE Transactions on Transportation Electrification*, 5(2):416–432, 2019.

[65] Naresh K. Pilli, M. Raghuram, Avneet Kumar, and Santosh K. Singh. Single dc-source-based seven-level boost inverter for electric vehicle application. *IET Power Electronics*, 12(13):3331–3339, 2019.

[66] Anton Kersten, Oskar Theliander, Emma Arfa Grunditz, Torbjorn Thiringer, and Massimo Bongiorno. Battery Loss and Stress Mitigation in a Cascaded H-Bridge Multilevel Inverter for Vehicle Traction Applications by Filter Capacitors. *IEEE Transactions on Transportation Electrification*, 5(3):659–671, 2019.

Appendices

A

Firing Angles for the EPM Scheme

TABLE A.1: Firing Sngles (in deg) for EPM Scheme from 7- to 31-Level

MLI Levels	α_1	α_2	α_3	α_4	α_5	α_6	α_7	α_8	α_9	α_{10}	α_{11}	α_{12}	α_{13}	α_{14}	α_{15}
7-level	25.71	51.43	77.14												
9-level	20.00	40.00	60.00	80.00											
11-level	16.36	32.72	49.09	65.45	81.81										
13-level	13.85	27.69	41.54	55.38	69.23	83.08									
15-level	12.00	24.00	36.00	48.00	60.00	72.00	84.00								
17-level	10.59	21.20	31.76	42.35	52.94	63.53	74.12	84.71							
19-level	9.47	18.90	28.42	37.89	47.37	56.84	66.32	75.79	85.26						
21-level	8.57	17.14	25.71	34.29	42.86	51.43	60.00	68.57	77.14	85.71					
23-level	7.83	15.70	23.49	31.30	39.13	46.96	54.78	62.61	70.43	78.26	86.09				
25-level	7.20	14.40	21.60	28.80	36.00	43.20	50.40	57.60	64.80	72.00	79.20	86.40			
27-level	6.67	13.33	20.00	26.67	33.33	40.00	46.67	53.33	60	66.67	73.33	80.00	86.67		
29-level	6.21	12.41	18.62	24.83	31.03	37.24	43.45	49.66	55.86	62.07	68.28	74.48	80.69	86.90	
31-level	5.81	11.61	17.42	23.23	29.03	34.84	40.65	46.45	52.26	58.06	63.87	69.68	75.48	81.29	87.10

B

Firing Angles for the HEPM Scheme

TABLE B.1: Firing Angles (in deg) for HEPM Scheme from 7-Level to 31-Level

MLI Levels	α_1	α_2	α_3	α_4	α_5	α_6	α_7	α_8	α_9	α_{10}	α_{11}	α_{12}	α_{13}	α_{14}	α_{15}
7-level	22.50	45.00	67.50												
9-level	18.00	36.00	54.00	72.00											
11-level	15.00	30.00	45.00	60.00	75.00										
13-level	12.86	25.71	38.57	51.43	64.29	77.14									
15-level	11.25	22.50	33.75	45.00	56.25	67.50	78.75								
17-level	10.00	20.00	30.00	40.00	50.00	60.00	70.00	80.00							
19-level	9.00	18.00	27.00	36.00	45.00	54.00	63.00	72.00	81.00						
21-level	8.18	16.36	24.55	32.73	40.91	49.09	57.27	65.24	73.63	81.82					
23-level	7.50	15.00	22.50	30.00	37.50	45.00	52.50	60.00	67.50	75.00	82.50				
25-level	6.92	13.85	20.77	27.69	34.62	41.54	48.46	55.38	62.31	69.23	76.15	83.08			
27-level	6.43	12.86	19.29	25.72	32.15	38.58	45.00	51.43	57.86	64.29	70.71	77.14	83.57		
29-level	6.00	12.00	18.00	24.00	30.00	36.00	42.00	48.00	54.00	60.00	66.00	72.00	78.00	84.00	
31-level	5.63	11.25	16.88	22.50	28.13	33.75	39.38	45.00	50.63	56.25	61.88	67.5	73.13	78.75	84.38

C

Firing Angles for the SHE Scheme

TABLE C.1: Firing Angles (in deg) for SHE Scheme from 7- to 31-Level

MLI Levels	α_1	α_2	α_3	α_4	α_5	α_6	α_7	α_8	α_9	α_{10}	α_{11}	α_{12}	α_{13}	α_{14}	α_{15}
7-level	8.35	28.90	54.78												
9-level	7.34	21.72	36.76	60.20											
11-level	5.11	16.96	30.24	42.24	63.75										
13-level	5.27	14.70	22.75	36.84	46.13	66.28									
15-level	5.50	12.46	19.08	30.75	39.28	50.13	67.73								
17-level	2.16	10.34	19.58	24.60	33.46	43.98	52.25	69.45							
19-level	2.91	8.58	16.76	23.55	27.63	38.59	45.51	54.91	70.41						
21-level	1.38	8.08	15.40	20.00	27.06	31.08	42.19	46.95	57.18	71.16					
23-level	2.74	6.64	13.07	20.00	22.80	29.83	34.89	44.39	48.51	59.08	71.76				
25-level	1.90	6.35	11.85	17.69	22.21	26.21	32.00	38.57	45.70	50.16	60.66	72.25			
27-level	0.85	5.87	11.65	14.71	22.03	24.04	28.44	35.35	40.71	47.20	51.46	62.15	72.55		
29-level	3.60	3.82	10.64	14.80	18.91	23.66	25.91	32.51	35.98	44.40	47.06	53.70	63.02	73.04	
31-level	1.25	5.26	8.71	14.29	17.56	21.77	24.97	29.43	33.70	38.90	45.85	47.77	55.60	63.76	73.53

D

Firing Angles for the NLM Scheme

TABLE D.1: Firing Angles (in deg) for NLM Scheme from 7- to 31-Level

MLI Levels	α_1	α_2	α_3	α_4	α_5	α_6	α_7	α_8	α_9	α_{10}	α_{11}	α_{12}	α_{13}	α_{14}	α_{15}
7-level	9.60	30.00	56.44												
9-level	7.18	22.02	38.68	61.04											
11-level	5.74	17.46	30.00	44.43	64.19										
13-level	4.78	14.48	24.62	35.69	48.59	66.44									
15-level	4.10	12.37	20.92	30.00	40.01	51.79	68.21								
17-level	3.58	10.81	18.21	25.94	34.23	43.43	54.34	69.64							
19-level	3.18	9.59	16.13	22.89	30.00	37.67	46.24	56.44	70.81						
21-level	2.87	8.63	14.48	20.49	26.74	33.37	40.54	48.59	58.21	71.81					
23-level	2.61	7.84	13.14	18.55	24.15	30.00	36.22	42.99	50.60	59.73	72.66				
25-level	2.39	7.18	12.02	16.96	22.02	27.28	32.80	38.68	45.10	52.34	61.04	73.40			
27-level	2.20	6.63	11.09	15.62	20.25	25.03	30.00	35.23	40.83	46.95	53.87	62.20	74.06		
29-level	2.05	6.15	10.29	14.48	18.75	23.13	27.66	32.39	37.38	42.73	48.59	55.23	63.26	74.64	
31-level	1.91	5.74	9.59	13.49	17.46	21.51	25.68	30.00	34.52	39.30	44.43	50.06	56.44	64.16	75.16

E

Firing Angles for the mNLM Schemes

TABLE E.1: Firing Angles (in deg) for mNLM Scheme from 7- to 31-Level

MLI Levels	α_1	α_2	α_3	α_4	α_5	α_6	α_7	α_8	α_9	α_{10}	α_{11}	α_{12}	α_{13}	α_{14}	α_{15}
7-level	8.60	27.60	50.44												
9-level	6.80	20.80	36.30	55.90											
11-level	5.44	16.46	28.60	42.03	59.36										
13-level	4.58	13.98	23.62	34.19	46.29	62.04									
15-level	4.00	11.97	19.92	29.00	38.51	49.59	64.11								
17-level	3.48	10.51	17.71	24.94	33.13	41.93	52.14	65.64							
19-level	3.08	9.29	15.73	21.89	29.00	36.57	44.74	54.34	67.01						
21-level	2.77	8.43	14.18	19.99	25.74	32.37	39.44	47.19	56.21	68.21					
23-level	2.51	7.64	12.84	18.15	23.65	29.00	35.22	41.89	49.20	57.73	69.26				
25-level	2.39	7.08	11.82	16.66	21.62	26.78	32.10	37.68	44.10	51.04	59.24	70.30			
27-level	2.20	6.53	10.89	15.32	19.95	24.63	29.5	34.63	39.83	45.95	52.67	60.50	71.06		
29-level	2.05	6.05	10.09	14.28	18.45	22.73	27.26	31.89	36.78	41.73	47.59	54.03	61.63	71.84	
31-level	1.91	5.64	9.49	13.29	17.26	21.21	25.28	29.60	34.02	38.70	43.73	49.16	55.34	62.66	72.56

Index

Printed in the United States
by Baker & Taylor Publisher Services